Inverse and Ill-Posed Problems Series

Characterisation of Bio-Particles
from Light Scattering

Also available in the Inverse and Ill-Posed Problems Series:

Carleman Estimates for Coefficient Inverse Problems and Numerical Applications
M.V. Klibanov and A.A Timonov

Counterexamples in Optimal Control Theory
S.Ya. Serovaiskii

Inverse Problems of Mathematical Physics
M.M. Lavrentiev, A.V. Avdeev, M.M. Lavrentiev, Jr., and V.I. Priimenko

Ill-Posed Boundary-Value Problems
S.E. Temirbolat

Linear Sobolev Type Equations and Degenerate Semigroups of Operators
G.A. Sviridyuk and V.E. Fedorov

Ill-Posed and Non-Classical Problems of Mathematical Physics and Analysis
Editors: M.M. Lavrent'ev and S.I. Kabanikhin

Forward and Inverse Problems for Hyperbolic, Elliptic and Mixed Type Equations
A G. Megrabov

Nonclassical Linear Volterra Equations of the First Kind
A.S. Apartsyn

Poorly Visible Media in X-ray Tomography
D.S. Anikonov, V.G. Nazarov, and I.V. Prokhorov

Dynamical Inverse Problems of Distributed Systems
V.I. Maksimov

Theory of Linear Ill-Posed Problems and its Applications
V.K. Ivanov, V.V. Vasin and V.P. Tanana

Ill-Posed Internal Boundary Value Problems for the Biharmonic Equation
M.A. Atakhodzhaev

Investigation Methods for Inverse Problems
V.G. Romanov

Operator Theory. Nonclassical Problems
S.G. Pyatkov

Inverse Problems for Partial Differential Equations
Yu.Ya. Belov

Method of Spectral Mappings in the Inverse Problem Theory
V. Yurko

Theory of Linear Optimization
I.I. Eremin

Integral Geometry and Inverse Problems for Kinetic Equations
A.Kh. Amirov

Computer Modelling in Tomography and Ill-Posed Problems
M.M. Lavrent'ev, S.M. Zerkal and O.E. Trofimov

An Introduction to Identification Problems via Functional Analysis
A. Lorenzi

Coefficient Inverse Problems for Parabolic Type Equations and Their Application
P.G. Danilaev

Inverse Problems for Kinetic and Other Evolution Equations
Yu.E. Anikonov

Inverse Problems of Wave Processes
A.S. Blagoveshchenskii

Uniqueness Problems for Degenerating Equations and Nonclassical Problems
S.P. Shishatskii, A. Asanov and E.R. Atamanov

Uniqueness Questions in Reconstruction of Multidimensional Tomography-Type Projection Data
V.P. Golubyatnikov

Monte Carlo Method for Solving Inverse Problems of Radiation Transfer
V.S. Antyufeev

Introduction to the Theory of Inverse Problems
A.L. Bukhgeim

Identification Problems of Wave Phenomena - Theory and Numerics
S.I. Kabanikhin and A. Lorenzi

Inverse Problems of Electromagnetic Geophysical Fields
P.S. Martyshko

Composite Type Equations and Inverse Problems
A.I. Kozhanov

Inverse Problems of Vibrational Spectroscopy
A.G. Yagola, I.V. Kochikov, G.M. Kuramshina and Yu.A. Pentin

Elements of the Theory of Inverse Problems
A.M. Denisov

Volterra Equations and Inverse Problems
A.L. Bughgeim

Small Parameter Method in Multidimensional Inverse Problems
A.S. Barashkov

Regularization, Uniqueness and Existence of Volterra Equations of the First Kind
A. Asanov

Methods for Solution of Nonlinear Operator Equations
V.P. Tanana

Inverse and Ill-Posed Sources Problems
Yu.E. Anikonov, B.A. Bubnov and G.N. Erokhin

Methods for Solving Operator Equations
V.P. Tanana

Nonclassical and Inverse Problems for Pseudoparabolic Equations
A. Asanov and E.R. Atamanov

Formulas in Inverse and Ill-Posed Problems
Yu.E. Anikonov

Inverse Logarithmic Potential Problem
V.G. Cherednichenko

Multidimensional Inverse and Ill-Posed Problems for Differential Equations
Yu.E. Anikonov

Ill-Posed Problems with A Priori Information
V.V. Vasin and A.L. Ageev

Integral Geometry of Tensor Fields
V.A. Sharafutdinov

Inverse Problems for Maxwell's Equations
V.G. Romanov and S.I. Kabanikhin

INVERSE AND ILL-POSED PROBLEMS SERIES

Characterisation of Bio-Particles from Light Scattering

V.P. Maltsev and K.A. Semyanov

UTRECHT • BOSTON
2004

VSP
an imprint of Brill Academic Publishers
P.O. Box 346
3700 AH Zeist
The Netherlands

Tel: +31 30 692 5790
Fax: +31 30 693 2081
vsppub@brill.nl
www.brill.nl
www.vsppub.com

© Copyright 2004 by Koninklijke Brill NV, Leiden, The Netherlands.
Koninklijke Brill NV incorporates the imprints Brill Academic Publishers, Martinus Nijhoff Publishers and VSP.

First published in 2004

ISBN 90-6764-413-7

All rights reserved. No part of this publication may be reproduced, stored in a retrieval system, or transmitted in any form or by any means, electronic, mechanical, photocopying, recording or otherwise, without the prior permission of the copyright owner.

A C.I.P. record for this book is available from the Library of Congress

Printed in The Netherlands by Ridderprint bv, Ridderkerk.

Preface

The electromagnetic scattering by small particles is under significant interest in a great variety of science and technologies. For example, this subject is important to climatology remote sensing of the Earth and planetary atmospheres. Another example is bioparticles. Biological particles play an important role in environment involving into processes related to human health, ecology, biotechnology, etc. The most important biological particles such as blood cells, spores, microorganisms, viruses can be effectively detected by means of light scattering which provides a non-invasive, express, fully-automated analysis. Nevertheless the polymorphism and complexity of biological particles pose specific approaches in simulation, measurement, and interpretation of light-scattering data.

Electromagnetic scattering by particles was considered in detail in the well-known monographs published by van de Hulst (1957), Kerker (1969), and Bohren and Huffman (1983). However the authors basically considered the direct light-scattering problem, computation of light-scattering field from predefined particle characteristics. At present the modern light-scattering theories allow simulation of light scattering by an individual particle of arbitrary shape and structure (Mishchenko et al (2000)). The progress in development of improved analytical and numerical methods has opened ways in development of new methods for solution of the inverse light-scattering problem, determination of particle characteristics from light scattering. On the other hand the rapid advancement in computers and instrumental techniques over the past two decades has resulted in an increasing of amount and improvement of quality of light-scattering data measured experimentally. Merging light-scattering simulation and experimental technique gives a new chance in complete characterization of individual particles from light scattering, especially bioparticles.

Today, all larger hospitals have facilities for detailed screening and characterisation of bacterial and blood cells of patients. Also many specialised research laboratories in cell biology or immunology all over the world routinely apply cell characterisation and sorters to facilitate their research. The most important technique that is applied to this end is Flow Cytometry. A flow cytometer can identify single cells (at speeds as high as 50,000 cells per second) using Elastic Light Scattering from the cell, and fluorescence from fluorescent probes, which bind to specific molecules on the cell surface or in the cell. Several fluorescence signals are measured, in combination with light scattering signals. In this way a number of independent measurements are obtained for every cell in the sample, and this information is used to identify different subsets in the original cell sample. Unfortunately at present 95% sorting facility of flow cytometers is based on an analysis of fluorescence.

During the last decade researchers from the laboratory of Cytometry and Biokinetics, Institute of Chemical Kinetics and Combustion, Novosibirsk, Russian Federation have developed a revolutionary new approach towards measuring light scattering in flow cytometry (Chernyshev et al (1995)). This Scanning Flow Cytometer (SFC) allows detailed measurement of the full angular light scattering signal (i.e. much more than in standard Flow Cytometry, were light scattering is measured over only two different angles). The SFC offers exiting new ways to use flow cytometry, ranging from extending and improving routine measurements in haematology laboratories to applying it in completely new fields (such as microbiology). The framework for this book is formed by the scientific articles of the Siberian team.

Therefore, the primary aim of this treatise is to provide a systematic state-of-the-art summary of the light-scattering of bioparticles, including brief consideration of analytical and numerical methods for computing electromagnetic scattering by single particles, detailed discussion of the instrumental approach used in measurement of light scattering, analysis of the methods used in solution of the inverse light-scattering problem, introduction of the results dealing with practical analysis of biosamples. Considering the widespread need for this information in optics, remote sensing, engineering, medicine, and biology, we hope that the book will be useful to many graduate students, scientists, and engineers working on various aspects of electromagnetic scattering and its applications.

The volume is opened with a concise introduction of analytical and numerical methods used for computing electromagnetic scattering by individual particles ranging from the Mie theory to Discrete Dipole Approximation. The Wentzel-Kramer-Brillouin approximation and T-matrix method are considered in Chapter 1. Chapter 2 gives a detailed overview of optical set-ups of two generations of the Scanning Flow Cytometers. The transformation of polarizing properties of light caused by scattering by a particle and propagation through optical set-up of the SFC is described using the Mueller-matrix presentation. Substantial and important part of the book relates to introduction of methods solving the inverse light-scattering problem for individual particles (Chapter 3). We demonstrate the algorithms that allow us to determine characteristics of spherical particles, spherical particles with a cover, absorbing spherical particles, sub-micron spherical particles, prolate spheroids. These methods were applied to characterize a number of samples. The applications described in Chapter 4 show an applicability of experimental and theoretical approaches in analysis of polymer beads, polymer bispheres, milk fat particles, red blood cells, *E.coli* cells, blood platelets, lymphocytes.

We thank our colleagues from the laboratory of Cytometry and Biokinetics: Andrey Chernyshev, Peter Tarasov, Vycheslav Nekrasov, Ilya Skribunov, Maxim Yurkin, Alexey Zharinov, and Konstantin Gilev. Authors special acknowledge Alexander Shvalov and Ivan Surovtsev for their contribution to development of Scanning Flow Cytometry. This book and new optical technology would not appear without financial support from Siberian Branch of the Russian Academy of Science (integration grants N70-2000 and N115-2003), the Russian Foundation for Basic Research (grants 00-02-17467-a, 02-02-08120-inno, 03-04-48852-a), NATO "Science for Peace" programme (project 977976 – Flow Cytometry). We thankfully acknowledge this continuing partnership.

Contents

Introduction   1

Chapter 1. Direct light-scattering problem of individual particles   5
1.1. Mie theory 5
1.2. T-matrix approach 8
1.3. Discrete dipole approximation 9
1.4. Wentzel–Kramer–Brillouin approximation 13

Chapter 2. Flow cytometry in measurement of light scattering of individual particles   19
2.1. Flow cytometry in measurement of light scattering of individual particles 19
2.2. Scanning flow cytometer 20
2.3. Polarizing scanning flow cytometer 27

Chapter 3. Inverse light-scattering problem of individual particles   35
3.1. Homogeneous spherical particles 35
3.2. Spherical particles with absorption 55
3.3. Spherical particles with a cover 61
3.4. Spheroidal particles 65
3.5. Neural network 68

Chapter 4. Applications   **73**
4.1. Polymer particles 73
4.2. Polymer beads analyzed with polarizing scanning flow cytometer . 81
4.3. Polymerization 85
4.4. Milk fat particles 85
4.5. Red blood cells 87
4.6. Bacterial cells 106
4.7. Lymphocytes 114
4.8. Bispheres 117

Conclusion   **121**

Acknowledgements   **125**

Bibliography   **127**

Introduction

Light scattering is a complex phenomenon that has found many uses in bioparticle discrimination. Each of the cellular components that have a refractive index different from that of the surrounding medium scatters light out of the incident beam. First of all an analysis of a biological cell assumes a presence of water as a surrounded medium that leads to a negligibly small difference between refractive indices of cell and medium. This fact supposes a usage of well-known approximations in simulation of light scattering from biological particles. From the other hand complexity of biological cells in shapes and internal structure assumes usage of delicate effects in light scattering such as depolarization, optical activity, etc. In general the approximations do not provide a simulation of these light- scattering effects. Consequently in order to characterize a biological cell from light scattering, exact light-scattering methods should be applied for light-scattering simulation.

Fortunately, because of the enormous progress in light-scattering modelling and simulation that was obtained in the late 1990s, we now see a new generation of researchers that are interested again in studying the optics of biological particles. Moreover, there is a strong renewed interest in experimental techniques that allow for rapid characterization. It is important to note that in most existing light scattering experiments (used e.g. in industrial applications for particle sizing) the scattering of a suspension of particles is obtained, that is, an average of scattering of all particles in the sample. This drastically diminishes the opportunities for the inverse problem, i.e. getting the particle's properties from the measured light scattering. Various light-scattering methods have been used in scientific research and industrial technologies in order to characterize particles. Instrumentally these methods can be divided into the following categories: (1) methods based on analysis of light scattered by a particle suspension; and (2) methods based on analysis of light scattered by an individual particle. Flow cytometry may be considered to belong to the second category as the hydrofocusing system provides a means for measuring scattering and fluorescent signals of individual particles.

Flow cytometry has been developed into a very powerful and versatile technique. However, this has unfortunately never resulted in a well-organized study of light-scattering properties of biological particles, nor have any real 'killer applications' emerged, usually because other competing techniques were preferred. For instance, despite the impressive progress in the mid-1980s on the characterization of human white blood cells in Flow Cytometry using polarized light scattering, this technique never found widespread use because in Cytology the alternative of immunophenotyping using multiple fluorescence optical paths turned out to be a much more powerful technique. However this implies the necessity of preparing the cell samples, which can be a time consuming procedure, and off course implies the use of (expensive) chemicals for the fluorescent probes.

The scattered fields, including their state of polarization, can be detected as a function of scattering angle and/or wavelength. The resulting light-scattering data can in principle allow the partial identification or discrimination of a biological particle from the natural background. Light scattering from single particles is very sensitive to their morphology and optical properties. Therefore, light scattering can be a sensitive tool for discriminating between cell types or between healthy and malignant cells, or to probe changes in cells resulting from stimuli. Even better, in principle light scattering allows real characterisation of the cells (e.g. in terms of physical dimensions or optical properties) by an inversion of the measured light scattering signals. Such inverse scattering procedure would allow complete new applications of flow cytometry, or new avenues in routine applications. For instance, identifications that now need the use of expensive monoclonal antibodies may be possible on the pure (i.e. unstained) cells.

Useful information about particle parameters can be retrieved from the angular dependency of light-scattering intensity, a light-scattering profile (LSP). The LSP characterizes morphology of the particle. In investigation of morphological characteristics of particles (size, refractive index, shape, etc.), the LSP works like a fingerprint. Various parameters including angular locations of maxima and minima, maxima to minima ratios, etc., can be determined from the LSP and used for calculation of the particle characteristics. Measurements of the LSP of individual particles, however, require a special method for carrying and keeping the particle in the testing zone and normally such methods are based on optical or electrostatic tweezers or on hydrodynamic focusing flow systems. Various kinds of instruments employing an electrostatic holder for an individual particle were used for measurement of the entire angular dependency of light-scattering intensity using lasers (Wyatt (1972), Cooke and Kerker (1973), Marx and Mulholland (1983), Neukammer et al (2003)). A method using an optical trap to keep a particle in a fixed position was demonstrated (Doornbos et al (1996)). Instruments, in which a particle is kept in a fixed position, and a scanning detector or a plurality of detectors are used for recording the scattering, require a relatively long time for measurement of the light-scattering pattern of individual particles and do not allow accumulation of a sufficient amount of data for statistical analysis.

A Scanning Flow Cytometer (SFC) is an instrument designed to measure the angular light-scattering pattern of an individual particle (Maltsev et al (1996), Maltsev et al (1997), Soini et al (1998), Maltsev (2000)). The LSP is measured by guiding of a particle or a cell through the measurement cuvette of the SFC and illumination of the particle with laser light. At present the SFC allows measurement of LSPs of individual cells with a rate of 500 particles per second within the measurable range of scattering angles ranging from 5^0 to 100^0. Thus measurement of the experimental signal is possible, but a comparison of the measured scattering signal with theoretical models of the LSPs from arbitrarily shaped particles, such as cells, remains a problem. This problem is due to the complexity of models that describe the light-scattering properties of small particles when the wavelength of the incident light is comparable with the particle size.

Determination of particle parameters from the LSP requires an investigation of the complex inverse problem that is generally not available in closed analytical form. This is a challenging inverse light scattering problem and, not least, a demanding instrumentation problem. The main method that was used by most of the authors to determine the spherical particle parameters from light scattering is a fit of experimental LSPs to LSPs calculated from the Mie theory. This method requires significant computer time and a set of initial values of variable parameters of the fitting. Determination of parameters of nonspherical particles from the fitting is more complicated. Empirical or approximating solutions of inverse light-scattering problem would be appropriate to provide the real-time analysis of individual particles from light scattering (Patitsas (1973)). An empirical solution for the inverse-scattering problem of individual particles is provided by the optical strip-map technique introduced by Quist and Wyatt (1985). This technique was modified by Jones et al (1990) to determine the particle size distribution function and optical properties. The spectral method for determination of spherical particle parameters from the LSP was suggested by Ludlow and Everitt (1995). They represented the scattered field and irradiance by a Gegenbauer series, the Gegenbauer spectrum. The high-order cutoff of this spectrum can result directly in the determination of the size parameter of the particle. The authors anticipated the use of the spectrum in development of an algorithm to determine the particle refractive index as well. A fast Fourier transform technique was applied by Min and Gomez (1996) to size the particles of known refractive index from the LSP with an accuracy of 3%. The algorithm allowed the determination of the particle size with a rate of 20–30 kHz from the LSP recorded at angles ranging from 90° to 180°. Ulanowski et al (1998) applied neural networks to the inverse light scattering problem for spheres. Typical computation times required for training the networks and particle parameter recovery were 50 s and 50 ms, respectively, using MATLAB 4.2 that runs on a 166 MHz Intel 80586 processor. The parameter area covered by the networks was 0.5 μm $< d <$ 1.5 μm and $1.12 < m < 1.27$ for particle diameter d and relative refractive index m, respectively.

In order to study features and the behaviour of systems consisting of microparticles, it is important to use a method that provides a real-time system for determination of the size distribution of suspended particles. Absolute measurements of size distribution are applicable on a wide scale and can improve the understanding of results from these studies. The requirement of the flow cytometry in a high analytical rate directed us to develop a method for particle analysis which should provide a real-time determination of particle characteristics and should be independent on an instrument alignment. In this way, we tried to parameterize the inverse light-scattering problem where spherical particle characteristics are calculated from the LSP parameters with approximating equations. The parameterisation means a creation of approximating equations that relate LSP parameters to particle characteristics.

Development of parametric solution of the inverse light-scattering problem forwards us to study of performance of the neural network approach to determine characteristics of individual particles from light scattering. This approach was successfully applied to measure diameter and refractive index of polymer beads and sphered erythrocytes.

Chapter 1.

Direct light-scattering problem of individual particles

1.1. MIE THEORY

The theory for scattering by a dielectric sphere was developed independently by Lorentz in 1890 and Gustav Mie in 1908. The derivation of the solution is a straightforward application of classical electromagnetic theory. Mie theory requires the relative refractive index = refractive index of a particle/refractive index of a medium. One needs to know the refractive index of the particle m' (i.e., refractive index of the material of which the particle is composed).

The strategy of Mie solution of direct light-scattering problem for a spherical particle can be introduced with the following algorithm: the Maxwell equations are used to derive a wave equation for electromagnetic radiation in three dimensional space; this equation is expressed in spherical polar coordinates (r, θ, φ), with appropriate boundary conditions at the surface of the sphere. The result is a separable partial differential equation, the solution of which is expressed as an infinite series of products of orthogonal basis functions, including sins and cosines (for the dependence on φ), spherical Bessel functions (for the dependence on r), and associated Legendre polynomials (for the dependence on $\cos\theta$).

Seek a solution of a vector wave equation (which follows from the Maxwell's equations) for $\boldsymbol{E}(r)$

$$\Delta \boldsymbol{E} - \frac{\varepsilon\mu}{c^2}\frac{\partial^2 \boldsymbol{E}}{\partial t^2} = 0, \tag{1.1}$$

where c is the speed of light in vacuum, $m' = \sqrt{\varepsilon\mu}$. The boundary condition assumes that the tangential components of $\boldsymbol{E}(r)$ and $\boldsymbol{H}(r)$ be continuous across the surface of a particle. Assumption on the spherical surface of a particle allows solving the vector equation analytically considering a spherical coordinate system centred on a spherical particle. The solution can be expressed in the following terms:

$$S_1(\theta) = \sum_{n=1}^{\infty} \frac{2n+1}{n(n+1)} [a_n \pi_n(\cos\theta) + b_n \tau_n(\cos\theta)],$$
$$S_2(\theta) = \sum_{n=1}^{\infty} \frac{2n+1}{n(n+1)} [a_n \tau_n(\cos\theta) + b_n \pi_n(\cos\theta)] \tag{1.2}$$

where

$$\pi_n(\cos\theta) = \frac{1}{\sin\theta} P_n^1(\cos\theta), \qquad \tau_n(\cos\theta) = \frac{d}{d\theta} P_n^1(\cos\theta), \tag{1.3}$$

and where P_n^1 is the associated Legendre polynomial. The coefficients a_n and b_n are referred to as Mie scattering coefficients and are functions of size parameter $\alpha = \pi d m_0/\lambda$ and relative refractive index $m = m'/m_0$, where d is the particle diameter, m_0 is the refractive index of the medium, λ is the wavelength of incident light, and m' is the refractive index of the particle. The mathematical forms of these coefficients are given as ratios of Ricatti–Bessel functions:

$$a_n = -\frac{j_n(m\alpha)[\alpha j_n(\alpha)]' - [m\alpha j_n(m\alpha)]' j_n(\alpha)}{j_n(m\alpha)[\alpha h_n^{(1)}(\alpha)]' - [m\alpha j_n(m\alpha)]' h_n^{(1)}(\alpha)},$$
$$b_n = -\frac{m^2 j_n(m\alpha)[\alpha j_n(\alpha)]' - [m\alpha j_n(m\alpha)]' j_n(\alpha)}{m^2 j_n(m\alpha)[\alpha h_n^{(1)}(\alpha)]' - [m\alpha j_n(m\alpha)]' h_n^{(1)}(\alpha)}. \tag{1.4}$$

The primes in the scattering coefficients denote differentiation with respect to the arguments in brackets.

As a practical matter, one cannot actually compute an infinite sum (equation (1.2)); therefore it is always necessary to truncate the series and keep only enough terms to yield a sufficiently accurate approximation. Generally speaking, the required number of terms N is a little larger than α;

Chapter 1. Direct light-scattering problem of individual particles

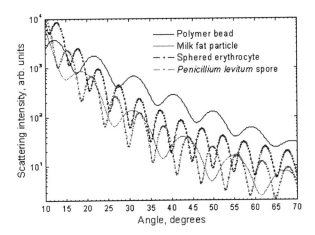

Figure 1.1. The light-scattering profiles, $0.5(|S_1|^2 + |S_2|^2)$, of a few particles simulated with Mie theory

the criterion developed by Bohren and Huffman (1983) based on extensive testing is that N should be the integer closest to $\alpha + 4\alpha^{1/3} + 2$. For a typical blood cell of 10 μm diameter and a visible wavelength of 0.5 μm, the size parameter $\alpha \approx 60$; thus the number of terms required in the summation is 78.

The Scattering matrix describes the relationship between incident and scattered electric field components perpendicular and parallel to the scattering plane as observed in the "far-field"

$$\begin{bmatrix} \boldsymbol{E}_{\|s} \\ \boldsymbol{E}_{\perp s} \end{bmatrix} = \frac{\exp\left(-ik(r-z)\right)}{-ikr} \begin{bmatrix} S_2 & S_3 \\ S_4 & S_1 \end{bmatrix} \begin{bmatrix} \boldsymbol{E}_{\|i} \\ \boldsymbol{E}_{\perp i} \end{bmatrix}. \quad (1.5)$$

For "far-field" observation of $\boldsymbol{E}_s$ at a distance r from a spherical particle of diameter d such that $r \gg d$, the scattering elements S_3 and S_4 equal zero. Hence for practical scattering experiments one measures intensity of the scattering light, $\boldsymbol{I}_s = \langle \boldsymbol{E}\boldsymbol{E}^* \rangle$, the above equation simplifies to the following:

$$\begin{bmatrix} \boldsymbol{I}_{\|s} \\ \boldsymbol{I}_{\perp s} \end{bmatrix} = \text{const} \begin{bmatrix} |S_2|^2 & 0 \\ 0 & |S_1|^2 \end{bmatrix} \begin{bmatrix} \boldsymbol{I}_{\|i} \\ \boldsymbol{I}_{\perp i} \end{bmatrix}. \quad (1.6)$$

To calculate $S_1(\theta)$ and $S_2(\theta)$ we intensively used in our work the algorithm described by Bohren and Huffman (1983). The program codes were developed in PASCAL and LabView environments. The Mie theory was applied to simulate of light scattering of polymer beads, sphered erythrocytes, and milk fat particles. The light-scattering profiles (angular dependency of light-scattering intensity, LSP) of these particles are shown in Figure 1.1. We applied the Mie theory calculation in experimental verification of the Scanning Flow Cytometer described in Chapter 2 that allows us to measure the LSPs of individual particles at angular interval ranging from 5° to 100°. The simulation of light scattering with Mie theory played significant role in development of parametric solution of the inverse light-scattering problem for spherical particles. Analysis of formation of the LSP structure under variation of particle characteristics was performed with Mie theory.

Additionally the algorithm described by Bohren and Huffman (1983) was used in simulation of light scattering of lymphocytes modelled by two concentric spheres. Moreover we have added the programming codes based on publication of Yang (2003) which allow us to calculate light scattering of multilayered sphere that models a lymphocyte scattering much better than two concentric spheres.

1.2. T-MATRIX APPROACH

At present, the -matrix approach is one of the most powerful and widely used tools for rigorously computing electromagnetic scattering by single and compounded nonspherical particles. Similar to Mie theory the T-matrix method allows the *exact* simulation of light scattering of single particles of basic shapes. However similar to Mie theory it is always necessary to truncate the infinite series and keep only enough terms to yield a sufficiently accurate *approximation*. This method can be sufficiently applied to simulate of light scattering of bacterial cells, blood platelets, cell aggregates, modelling these cells by prolate and oblate spheroids, bispheres, respectively. However, with non-symmetric and even slightly concave particles, such as mature red blood cells, the inherent numerical instabilities are encountered, which are currently under an attack using different quadratures and other variants of the T-matrix approach.

The strategy of the T-matrix solution of direct light-scattering problem can be introduced with the following algorithm: expanding the incident field in vector spherical wave functions regular at the origin; expanding the scattered field outside a circumscribing sphere of the scatterer in vector spherical wave functions regular at infinity; transforming the expansion coefficients of the incident field into those of the scattered field via T matrix

Chapter 1. Direct light-scattering problem of individual particles

and, if known, computing any scattering characteristic of a nonspherical particle. The method was first introduced by Waterman (1971) for single homogeneous scatterers. For spheres, all T-matrix method formulas reduce to those of the standard Mie theory.

The detailed review of the -matrix approach was introduced in the collective monograph (Mishchenko et al (2000)). Accordingly, almost all existing computer codes assume rotationally symmetric shapes both smooth, for example, spheroids and so-called Chebyshev particles, and sharp edged, for example, finite circular cylinders. Recent work has demonstrated the practical applicability of T-matrix method to particles without axial symmetry, for example, general ellipsoids, cubes, and clusters of spheres, although the computational complexity of these calculations is significantly greater than that for rotationally symmetric scatterers (Wielaard et al (1997)).

The elements of the -matrix are independent of the incident and scattered fields and depend only on the shape, size parameter, and refractive index of the scattering particle and on its orientation with respect to the reference frame, so that the matrix need be computed only once and then can be used in computations for any directions of light incidence and scattering. The public-domain T-matrix codes (http://www.giss.nasa.gov/crmim/) allow one to calculate the 2×2 amplitude matrix (equation (1.5)). Contrary to spherical particles the elements of the amplitude matrix depends on orientation of a particle, azimuth and polar angles measure out direction of incident laser beam, $S_i = S_i(\theta, \varphi)$. Moreover the elements S_3 and S_4 do not equal zero for non-spherical particles.

In our work we applied the T-matrix method to simulate light scattering of bacteria-like particles modelled by a prolate spheroid, cylinder. We expanded the public-domain T-matrix codes on a rod, a cylinder capped with hemispheres of the same radius as the cylinder. A rod is a most suitable model for rod-like bacterial cells, for instance, *Escherichia coli* cells. The light scattering of blood platelets was also computed with the T-matrix method modelling the cells by oblate spheroids. The LSPs of these particles calculated from T-matrix method are shown in Figure 1.2. The computing codes were written in LabView programming environment.

1.3. DISCRETE DIPOLE APPROXIMATION

The discrete dipole approximation (DDA) allows simulation of light scattering of the continuum target by a finite array of polarizable points. The points acquire dipole moments in response to the local electric field. The

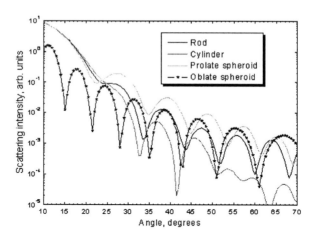

Figure 1.2. The light-scattering profiles, $0.5(|S_1|^2 + |S_2|^2 + |S_3|^2 + |S_4|^2)$, of a few particles simulated with the T-matrix method. The rod, cylinder, and prolate spheroid modelling a rod-like bacterium are equal volume. The oblate spheroid is a model of blood platelets. The symmetry axis of the particles coincides with direction of incident beam

dipoles interact between each other via their electric fields (Purcel and Pennypacker (1973), Draine (1988)) and DDA is also sometimes referred to as the coupled dipole approximation (Singham and Salzman (1986), Singham and Bohren (1987)). The theoretical basis for the DDA, including radiative reaction corrections, was summarized by Draine (1988).

For a finite array of point dipoles the scattering problem may be solved exactly, the approximation assumes a replacement of the continuum target by an array of N-point dipoles. The replacement requires specification of both the geometry (location r_j of the dipoles $j = 1, \ldots, N$) and the dipole polarizabilities α_j.

There has been some controversy concerning the best method of assigning the dipole polarizabilities. Purcel and Pennypacker (1973) used the Clausius-Mossotti polarizabilities:

$$\alpha_j^{CM} = \frac{3d^3}{4\pi} \frac{\varepsilon_j - 1}{\varepsilon_j + 2}, \qquad (1.7)$$

where d is the size of sub-volume and ε_j is the dielectric permeability of the target material at location r_j. According to Jackson (1975) the Clausius-

Mossotti prescription is exact in the limit $kd \to 0$ for an infinite cubic lattice, where k is the wavenumber. Draine (1988) and Goedecke and O'Brien (1988) showed that the polarizabilities should also include a radiative reaction correction of $O[(kd)^3]$. Draine and Goodman (1993) proposed the lattice dispersion prescription (LDR) for dipole polarizabilities. In the long-wavelength limit $kd \ll 1$, the polarizability $\alpha(\omega)$ is given as a series expansion in powers of kd and $m^2 = \varepsilon$:

$$\alpha^{\mathrm{LDR}} = \frac{\alpha^{\mathrm{CM}}}{1 + (\alpha^{\mathrm{CM}}/d^3)[(b_1 + m^2 b_2 + m^2 b_3 S)(kd)^2 - (2/3)i(kd)^3]}, \quad (1.8)$$

where $b_1 = -1.891531$, $b_2 = 0.1648469$, $b_3 = -1.7700004$, $S = \sum_{j=1}^{3}(\hat{a}_j \hat{e}_j)^2$, and $\hat{a}$ and $\hat{e}$ are unit vectors defining the incident direction and the polarization state. The LDR prescription for $\alpha(\omega)$ is, by construction, optimal for wave propagation on an infinite lattice, and it is reasonable to assume that it will also be a good choice for finite dipole arrays. Extensive DDA calculations for spheres, comparing different prescriptions for the dipole polarizability, confirm that the LDR prescription appears to be best for $|m|kd < 1$ (Draine and Goodman (1993)).

The direct electromagnetic scattering problem must be solved for the target array of point dipoles ($j = 1, \ldots, N$) with polarizabilities α_j, located at positions $\boldsymbol{r}_j$. Each dipole has a polarization $\boldsymbol{P}_j = \alpha_j \boldsymbol{E}_j$, where $\boldsymbol{E}_j$ is the electric field at $\boldsymbol{r}_j$ that is due to the incident wave $\boldsymbol{E}_{\mathrm{inc},j} = \boldsymbol{E}_0 \exp(i\boldsymbol{k}\cdot\boldsymbol{r}_j - i\omega t)$ plus the contribution of each of the other $N-1$ dipoles:

$$\boldsymbol{E}_j = \boldsymbol{E}_{\mathrm{inc},j} - \sum_{k \neq j} A_{jk} \boldsymbol{P}_k, \quad (1.9)$$

where $A_{jk} \boldsymbol{P}_k$ is the electric field at $\boldsymbol{r}_j$ that is due to dipole $\boldsymbol{P}_k$ at location $\boldsymbol{r}_k$, including retardation effects. Each element A_{jk} is a 3×3 matrix:

$$A_{jk} = \frac{\exp(i k r_{jk})}{r_{jk}} \left(k^2 (\hat{\boldsymbol{r}}_{jk} \hat{\boldsymbol{r}}_{jk} - I_3) + \frac{i k r_{jk} - 1}{r_{jk}^2} (3 \hat{\boldsymbol{r}}_{jk} \hat{\boldsymbol{r}}_{jk} - I_3) \right), \quad (1.10)$$

$$j \neq k,$$

where $k \equiv \omega/c$, $r_{jk} = |\boldsymbol{r}_j - \boldsymbol{r}_k|$, $\hat{\boldsymbol{r}}_{jk} = (\boldsymbol{r}_j - \boldsymbol{r}_k)/r_{jk}$, and I_3 is the 3×3 identity matrix. Defining $A_{jj} = \alpha_j^{-1}$ reduces the scattering problem to finding the polarizations $\boldsymbol{P}_j$ that satisfy a system of $3N$ complex linear equations:

$$\boldsymbol{E}_{\mathrm{inc},j} = \sum_{k=1}^{N} A_{jk} \boldsymbol{P}_k. \quad (1.11)$$

Differential scattering cross sections may also be directly evaluated once the P_j are known (Draine and Flatau (1994)). In the far field the scattered electric field is given by:

$$\boldsymbol{E}_{\text{sca}} = \frac{k^2 \exp(ikr)}{r} \sum_{j=1}^{N} \exp(-ik\hat{r} \cdot \boldsymbol{r})(\hat{r}\hat{r} - I_3)\boldsymbol{P}_j. \qquad (1.12)$$

Rather than direct methods for solving equation (1.11), CCG methods for finding P iteratively have proven effective and efficient. As P has $3N$ unknown elements, CCG methods in general are only guaranteed to converge in $3N$ iterations. In fact, however, $\sim 10-10^2$ iterations are often found to be sufficient to obtain a solution to high accuracy (Draine (1988), Goodman et al (1991)). The choice of conjugate-gradient variant may influence the convergence rate (Sarkar et al (1988), Peterson et al (1991), Freund and Nachtigal (1991)).

The computational burden in the CCG method primarily consists of matrix-vector multiplications of the form $A\boldsymbol{v}$. Goodman et al (1991) showed that the structure of the matrix A implies that the multiplications are essentially convolutions and FFT methods can be employed to evaluate $A \cdot \boldsymbol{v}$ in $O(N \ln N)$ operations rather than in the $O(N^2)$ operations that are required for general matrix-vector multiplication. Since N is large this is an important calculational breakthrough. FFT methods require that the dipoles be situated on a periodic lattice, which is most simply taken to be cubic.

If the N dipoles were located at arbitrary positions $\boldsymbol{r}_j$, then the $9N^2$ elements of A would be nondegenerate. Storage of these elements (with 8 bytes/complex number) would require $72(N/10^3)^2$ Mbytes. By locating the dipoles on a lattice, the elements of A become highly degenerate, since they depend only on the displacement $\boldsymbol{r}_j - \boldsymbol{r}_i$. As a result the memory requirements depend approximately linearly on N rather than on N^2.

As with all numerical simulations of wave phenomena, the maximum size of the discretization is bounded by the wavelength ($d < \lambda/10$). This bound results, for particle in resonance region, in DDA models containing up to 10^8 dipoles. We therefore see that memory requirements begin to be a consideration for targets that are in resonance region. The computer processor requirements also are significant for large targets.

Hoekstra et al (1998) have parallelized the DDA and showed that it runs very efficiently on distributed memory computers, provided that the number of dipoles per processor is large enough. They run DDA simulations up to $7 \cdot 10^6$ dipoles, only limited by the available amount of memory.

Chapter 1. Direct light-scattering problem of individual particles 13

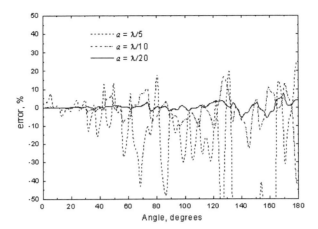

Figure 1.3. The relative errors in DDA simulations for different discretization size. The DDA simulations were compared with the exact solutions of Mie theory for sphere with diameter $d = 4.56$ μm and relative index of refraction $m = 1.10$

The size of sub-volume a must be small enough to ensure that response to electromagnetic field is the response of an ideal induced dipole. The size should be in range $\lambda/20 < d < \lambda/10$ (Hoekstra and Sloot (1993)). But in some cases we can use even $d \leq \lambda/5$. Using the parallel DDA code produced by Hoekstra et al (1998) we compared the DDA simulations with the Mie calculation for homogeneous sphere with a diameter of 4.56 μm and a relative refractive index of 1.10; the wavelength was 0.6328 μm. The relative errors of DDA simulations as a function of the scattering angle are presented in Figure 1.3 for three different discretization size $\lambda/5$, $\lambda/10$, $\lambda/20$. The relative error for discretization size of $\lambda/5$ is in range of 10 % for the scattering angles before 55 degrees. This is quite enough for the comparisons with the experimental curves in a range of scattering angle from 10 to 55 degrees.

1.4. WENTZEL–KRAMER–BRILLOUIN APPROXIMATION

In order to analyse biological suspensions with an optical technique, approximation methods are widely used to describe the light-scattering pattern. These approximations usually assume that the relative refractive index m of particles is close to 1; thus, the particles are called 'optically soft'. The most commonly used light-scattering approximations are as follows: Rayleigh's scattering, Rayleigh–Gans–Debye approximation (RGD), anomalous diffraction (AD), and Fraunhofer diffraction (FD) (van de Hulst (1957), Kerker (1969)). The most general and simple solution of the light-scattering problem for 'optically soft particles' can be produced on the basis of the integral wave equation in the Wentzel–Kramers–Brillouin (WKB) approximation (Klett et al. (1992), Lopatin and Shepilevich (1996)).

The solution of the integral wave equation allows a presentation of the amplitude of the scattered field in the far zone that is as follows:

$$\boldsymbol{E}^S(\boldsymbol{r}) = f(\boldsymbol{o}, \boldsymbol{i}) \frac{e^{ikR}}{R},$$

$$f(\boldsymbol{o}, \boldsymbol{i}) = \frac{k^2}{4\pi} \int_{V'} \{-\boldsymbol{o} \times [\boldsymbol{o} \times \boldsymbol{E}(\boldsymbol{r}')]\}[m^2(\boldsymbol{r}') - 1] \exp\left[-ik(\boldsymbol{r}'\boldsymbol{o})\right] dV',$$

(1.13)

where $kR \gg 1$, $k = 2\pi/\lambda$ is a wavevector of the particle substance, R is the distance from the particle to the detection point along the scattering direction, V' is a volume of the scatterer, $\boldsymbol{i}$ and are unit vectors in the direction of the propagation of the incident and scattered light, respectively, and $\boldsymbol{E}(\boldsymbol{r}')$ is a time-independent part of the electric field within the particle.

Expression (1.13) is an exact integral presentation of the amplitude of the scattered field in terms of the internal field $\boldsymbol{E}(\boldsymbol{r}')$. The WKB approximation has been obtained by substituting the exact field $\boldsymbol{E}(\boldsymbol{r}')$ by a plane wave with a wavevector which corresponds to the optical properties of the particle material. The direction and amplitude of the wave are not changed inside the scatterer. Thus, the internal field can be expressed as follows:

$$\boldsymbol{E}(\boldsymbol{r}') = \boldsymbol{e}_i \exp\left\{ik(\boldsymbol{r}'\boldsymbol{i}) + ik\int_{Z_1}^{Z'} m(z')\, dz'\right\}, \quad (1.14)$$

where $\boldsymbol{e}_i$ is a unit polarization vector of the incident light, $Z_1 = (\boldsymbol{r}_1 \cdot \boldsymbol{i})$ is a coordinate of the point $\boldsymbol{r}_1$ where the plane wave intersects the particle surface, and $Z' = (\boldsymbol{r}' \cdot \boldsymbol{i})$.

Substitution of (1.14) into (1.13) gives

$$f(\boldsymbol{o}, \boldsymbol{i}) = \frac{k^2}{4\pi}\{-\boldsymbol{o} \times [\boldsymbol{o} \times \boldsymbol{e}_i]\}V F(\boldsymbol{o}, \boldsymbol{i}),$$

$$F(\boldsymbol{o}, \boldsymbol{i}) = \frac{1}{V}\int_V [m^2 - 1]\exp\left(ik_s\boldsymbol{r}'\right)\exp\left\{ik\int_{Z_1}^{Z'}(m-1)\,dz'\right\}dV',$$

(1.15)

where $\boldsymbol{k}_s = k\boldsymbol{i}_s = k(\boldsymbol{i} - \boldsymbol{o})$ is directed along a bisector of the complementary scattering angle, $|\boldsymbol{i}_s| = 2\sin(\theta/2)$, θ is a scattering angle, i.e. the angle between $\boldsymbol{i}$ and $\boldsymbol{o}$.

Somewhat more precise results with a broader range of applicability can be obtained by use of the two-wave WKB approximation, a modification of the WKB approximation. With the two-wave WKB approximation the internal field is represented as a superposition of the incident wave and the wave propagating backward, reflected by the rear surface of the particle z'. The phase and the amplitude of the reflected wave are calculated, as in the case of normal reflection, from the plane boundary of the two dielectrics. The normalized internal electric field is as follows:

$$\boldsymbol{E} = \boldsymbol{E}^i \frac{2}{m+1} \exp\left[-i(m-1)\boldsymbol{k}\boldsymbol{r}_1\right][\exp\left(im\boldsymbol{k}\boldsymbol{r}\right) + R\exp\left(i(\alpha_1 - m\boldsymbol{k}\boldsymbol{r})\right)], \tag{1.16}$$

where $R = (1-m)/(1+m)$ is the reflection coefficient and $\chi_1 = 2mk(r'-r_2)$ is the phase shift of the backward wave at point r'.

In the case of a homogeneous sphere and for "optically soft particles", $(m^2 - 1) \approx 2(m - 1)$, expression (1.15) is transformed into the following:

$$|f(\boldsymbol{o}, \boldsymbol{i})| = \sin \chi \frac{k^2}{2\pi} (m-1)|F(\theta)|, \tag{1.17}$$

where

$$F(\theta) = \frac{4\pi a^2}{k_3} \int_0^1 J_0(\alpha \sin \theta \sqrt{1-t^2}) \sin[\alpha(m - \cos \theta)t] \exp\left(i\frac{\rho}{2}t\right) t \, dt,$$

where χ is an angle between $\boldsymbol{e}$ and $\boldsymbol{o}$, $k_3 = k(m - \cos \theta)$, a is radius of the sphere, $J_0(x)$ is the Bessel function of the zero order.

In order to analyse the extreme locations for WKB approximation we used equation (1.17) for calculations of light-scattering profile (LSP) minima. The locations of the first eight LSP minima expressed in terms of $2\alpha \sin(\theta/2)$ as a function of the phase-shift parameter ρ are shown in Figure 1.4 for both $m = 1.025$ and $m = 1.2$. We calculated the locations of the first eight minima for the fixed ρ starting from the $\theta = 0$. The functions reveal a similarity for both relative refractive indices that bound a range of the "optically soft particles", above the dashed line in Figure 1.4. Moreover, the functions for both the refractive indices are congruent at small angles and are independent of the relative refractive index of particles, below the dashed line in Figure 1.4. The locations of the minima that correspond to the RGD approximation and FD with asymptotes $\rho \ll 1$ and $\rho \gg 1$, respectively, are shown in Figure 1.4 as dotted and solid lines, respectively.

We determined the location θ_s of the extremum defined as a source of extrema. The source of extrema means that an extremum appears in

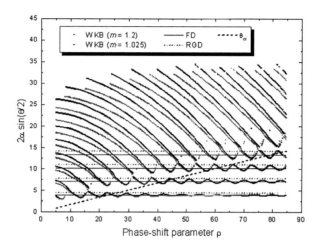

Figure 1.4. Locations of the light-scattering profile minima as a function of the phase-shift parameter ρ calculated from the WKB, RGD, and FD approximations. Dashed line presents the critical angle θ_{cr}

this point for certain particle parameters and then migrates to the forward direction with increasing α or ρ. The start of migration corresponds to a conversion of the minimum (maximum) into the maximum (minimum) in this point. This location must be independent of α, m, ρ and results from the condition for the function of $F(\theta)$ (equation (1.17)):

$$\frac{\partial F}{\partial \theta} \equiv 0 \qquad (1.18)$$

that provides the following set of equations:

$$\frac{\partial k_3}{\partial \theta} = k\sin\theta = 0,$$

$$\frac{\partial}{\partial \theta} J_0(\alpha \sin\theta \sqrt{1-t^2}) = -J_1(\alpha \sin\theta \sqrt{1-t^2})\alpha \cos\theta \sqrt{1-t^2} = 0, \quad (1.19)$$

$$\frac{\partial}{\partial \theta} sin\left[\alpha(m - t\cos\theta)\right] = \cos\left[[\alpha(m - t\cos\theta)]\alpha \sin\theta = 0.\right.$$

$\theta = 0$ and $\theta = \pi$ are solutions of the equation set (1.19) but $\theta = 0$ is an absolute maximum. Hence, $\theta_s = \pi$ is the source of LSP extrema for the WKB approximation and coincides with that for the RGD approximation of a homogeneous sphere. The locations of the minima expressed in terms of $2\alpha \sin(\theta/2)$ are monotonically reduced after an appearance in the source of extrema, whereas distances between the adjacent extrema are increased when α is increased (Figure 1.4). The following migration rule is inherent to both the WKB and RGD approximations: the extremum appears and moves to small scattering angles with an increasing value of α or m and, hence, ρ. This fact is confirmed from Figure 1.4 where the asymptotes of the minima location lines of the WKB calculations coincide with the dotted line of the RGD calculations.

The solid line of the FD calculations is different from both the WKB and RGD calculations. Relating to the large particles that satisfy the FD we can use the condition $\alpha \gg 1$ to transform equation (1.17) to the following:

$$\frac{|f(\theta = \pi)|^2}{\pi a^2} = \frac{(m-1)^2}{\pi} \left[\frac{\cos[\alpha(m+1)]}{(m+1)}\right]^2 + O(\alpha^{-1}) \qquad (1.20)$$

and, hence, the extrema appear with a period of π relating to the argument $\alpha(m+1)$. In contrast to the RGD approximation where $z = 2\alpha$ at the scattering angle π, the source of extrema θ_s expressed in terms of z must be divided by $(m+1)/2$. This conclusion is satisfied with an arbitrary value of α. Consequently, the locations of extrema expressed in terms of $2\alpha \sin(\theta/2)$ is reduced by $(m+1)/2$ for the FD calculation in relation to the RGD approximation (Figure 1.4).

Moreover, the results presented in Figure 1.4 allow us to define the critical angle θ_{cr} calculated from the following equation:

$$\theta_{cr} = 0.336(m-1). \qquad (1.21)$$

Equation (1.21) is an empirical equation of the dashed line (Figure 1.4) that divides the region into two parts: dipole approximation and diffraction. Equation (1.21) determines the angular interval $(\theta \leq \theta_{cr})$ where Fraunhofer diffraction can be applied.

With this analysis we have shown that the distance between minima is insensitive to a variation of particle refractive index. This fact proves the usage of this LSP parameter in the parametric solution of inverse light scattering problem (Section 3.1.2). The important result of this study is the empirical equation for the critical angle that is the upper boundary of the angular region of FD applicability. The other empirical equation has allowed an extension of well known formula of extrema locations retrieved from the RGD approximation (van de Hulst (1957)). These equations can be applied to calculate the extrema locations for angles above the critical angle.

Chapter 2.

Flow cytometry in measurement of light scattering of individual particles

2.1. FLOW CYTOMETRY IN MEASUREMENT OF LIGHT SCATTERING OF INDIVIDUAL PARTICLES

An approach for the measurement of the entire angular dependency of scattering intensity (light-scattering profile, LSP) that uses the motion of a particle carried by a stream was described by Loken et al (1976). An optical system recorded the scattering intensity versus time, which could then be related to intensity versus scattering angle, to obtain single-particle-scattering measurements for polar angles from 1° to 49°. To correct the measured scattering function for variations in illumination intensity and collection aperture as a function of particle position, the transfer function was determined empirically by introducing fluorescent particles into the flow.

A number of articles dealing with analysis of individual particles from light scattering were published in a special issue with introductory articles (Damaschke et al (1998), Kaye (1998)). Ludlow and Kaye (1979) used a scanning diffractometer with a single photomultiplier to measure the LSPs of polystyrene particles and spores. A rotating disk and 174 optical light guides permitted the measurement of scattering intensities of single particles at polar angles from 3° to 177° in 2.8 ms. To account for differences the optical properties of the guides, a special digital–analog converter in multiplication mode provided a correction to the analog signal. The design of the laser light-scattering instrument was developed by Kaye and Hirst with co-workers to measure the spatial distribution of light scattered by individual particles in a flow. This instrument was equipped with an intensified charge coupled device camera and recorded scattering profiles with a rate of 3 images/s at scattering angles from 5° to 30° (Kaye et al (1997)) and from 28° to 141° (Hirst and Kaye (1996)) to the beam axis and throughout 360° of azimuth.

Constantinides et al (1998) combined three-dimensional boundary-element analysis of EM wave scattering by penetrable particles and a specially developed fast pattern recognition technique to develop a new, low-cost apparatus for nonspherical axisymmetric particle analysis. The new apparatus allowed determination of size, shape, and refractive index of particles in liquid suspensions. Experimental results of light scattering by a mixture of spherical latex particles and glass beads, RBC's, or bacteria were in a good agreement with the respective numerical predictions. The fast pattern-recognition algorithm that was developed for the purpose of determining size, shape, and refractive index of the particles in a few seconds from a standard least-squares minimization algorithm.

Neukammer et al (2003) reported on two-dimensional angular-resolved light scattering of single blood cells, in particular, on sphered red blood cells (erythrocytes), on native erythrocytes elongated by hydrodynamic forces, and on white blood cells (lymphocytes). They used a laser flow cytometer together with an intensified CCD camera as a detector to measure differential light-scattering cross sections. By the same technique angular-resolved light scattering of an oriented agglomerate (dumbbell) consisting of two identical polystyrene microspheres was recorded.

2.2. SCANNING FLOW CYTOMETER

In this chapter the instrumental approach to measurement of light scattering of individual particles developed by the authors with co-workers is described. The technology defined as scanning flow cytometry was introduced in middle of 90th last century (Maltsev (1994), Chernyshev et al (1995)). The main aim of this technology is characterisation of individual particles from light scattering with a typical rate of a few hundred particles per second. Characterization means determination of physical particle characteristics like a size, shape, density etc. by means of solution of the inverse light-scattering (ILS) problem. In order to support the solution by sufficient amount of light-

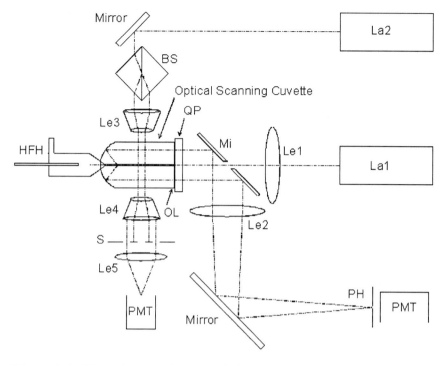

Figure 2.1. The schematic lay-out of the second generation of the SFC's optical system

scattering data we introduced a new generation of flow cytometers, Scanning Flow Cytometer (SFC). Contrary to the ordinary flow cytometry where the forward and side scattering signals are used to analyze the particles, the SFC allows the measurement of LSP. The theoretical and experimental sides of this technology are based on a solution of the ILS problem and SFC, respectively. At present, the SFC allows the measurement of the LSP of individual particles at angles ranging from 5° to 100° at a rate of 500 particles/s. The basic principle of the SFC was patented (Maltsev and Chernyshev (1997)).

2.2.1. Optical set-up

The schematic layout of the Scanning Flow Cytometer optical system is presented in Figure 2.1. Two He-Ne lasers are used to illuminate the particles within the cuvette. The coaxial laser La1 is used for generation of the scattering pattern and the orthogonal laser La2 is used for flow speed control and as a reference point in a time to angle transformation that will be described below. The beam of laser La1 is directed coaxially with the stream by a

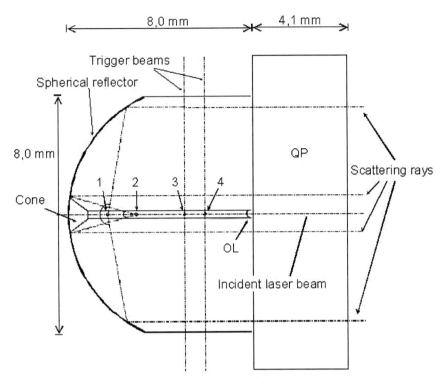

Figure 2.2. A detailed functional drawing of the optical cuvette

lens Le1 through a hole in the mirror Mi. The hydrodynamic focusing head (HFH in Figure 2.1) produces two concentric fluid streams: a sheath stream without particles and a probe stream that carries the analyzed particles. A differential pressure regulator is used for sample flow control. The fluidics system provides a probe stream with 12 μm diameter. The hydrodynamically focused stream is fed into the cuvette at the spherical end through a cone where the hydrodynamic focusing head is connected by mechanical pressure. The flow channel is truncated in the flat end of the cuvette where it is connected to an orthogonal outlet (OL). For this purpose the cuvette is combined with a quartz plate (QP). The 0.5 mm wide outlet OL is grooved at the flat end surface of the cuvette.

The optical function of the cuvette is shown in Figure 2.2. The laser beam La1 is focused to the cuvette using a lens Le1. The length of the laser beam waist is approximately 3 mm and the "sensing zone" lies between points 1 and 3 in Figure 2.2. The sample is fed into the cuvette from the spherical mirror side. The incident light beam is scattered from particles and reflected from the spherical reflector surface out from the cuvette. The scattered light is then reflected by the mirror Mi outside the cuvette and focused by a lens Le2 ($f = 300$ mm) into a pinhole PH ($d = 200$ μm) in front of a photomultiplier tube (PMT).

For each particle position within the sensing zone, there is a specific light scattering angle θ in which the rays are reflected in parallel to the stream axis by the spherical mirror. For instance, particle locations 1 and 2 correspond to scattering angles 100° and 15°, respectively (Figure 2.2). The specific angle of scattered light, which is reflected in parallel to the stream axis, changes continuously as the particle moves within the sensing zone. Only those rays reflected by the mirror and which are parallel to the stream axis are focused onto the pinhole PH and detected by the PMT. Thus, the LSP signal, which expresses the light scattering intensity at different scattering angles, is obtained by measuring the signal intensity as a function of time. The system is equipped with the laser La2 which is orthogonal to the probe stream. Using a beamsplitter BS, the orthogonal laser beam is split into two beams that are focused by an objective lens Le3 (NA = 0.2) into the probe stream at a 90° angle. The particles that cross the beams scatter light. The scattered light is collected by an objective lens Le4 (NA = 0.2) and a lens Le5 ($f = 30$ mm) and is detected by a PMT. A mask S with two beamstops is used to stop the direct illumination of the PMT. The two consecutive signals from the PMT are used to control the speed of the flow. The trigger signal controlling the AD-converter can be taken either from the orthogonal laser beam, or the LSP signal itself can be used for triggering. In the first case, the ADC is operated continuously and the AD-conversion is interrupted for read-out when a particle crosses the first orthogonal beam (point 3 in Figure 2.2).

2.2.2. Time to angle transformation

During a particle's motion within the sensing zone, the signal amplitude versus time can be directly related to the scattered light intensity versus scattering angle. The pinhole in the front of the PMT allows only the rays parallel to the probe stream to be detected. The location of a particle within the sensing zone, i.e., the distance l from the particle to the bottom of the spherical reflector, can be expressed as follows:

$$\delta = \cos^{-1}\left(\frac{m_0}{m_1}\cos\theta\right), \qquad (2.1)$$

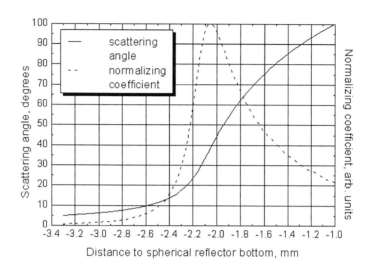

Figure 2.3. A graphical presentation of the function that relates the scattering angle θ to the location of the particle l (solid curve). A graphical presentation of the instrument function for normalizing the amplitude of measurement data (dashed curve)

$$l = R\left(1 - \cos\frac{\delta}{2}\right) + \left(R \sin\frac{\delta}{2} - \frac{d}{2}\right) \cot \delta + \frac{d}{2} \cot \theta, \qquad (2.2)$$

where θ is the scattering angle, m_0 is the refractive index of the liquid medium in the flow channel, m_1 is the refractive index of quartz, d is the diameter of the channel, and R is the radius of the spherical reflector. With the current setup the function that relates the scattering angle θ to the location of a particle l is an inverse function of equation (2.2) and its numerical presentation is shown in Figure 2.3 (solid curve).

The signal obtained from the ADC, the SFC trace, is not representative of a real LSP signal because it is modified by the instrument function. First, the light collection efficiency and the angular resolution are strongly dependent on a particle's location. In order to normalize the observed signal amplitudes, a numerical normalizing function was calculated point by point according to the real geometrical dimensions of the optical system. This function is presented in Figure 2.3 (dashed curve). Second, as the detection system records the scattering intensity as a function of time, a time to scattering angle transformation of the recorded signal must be performed. For this purpose the location of the particle within the sensing zone must be determined as a function of time and the location of the particle can be calculated using the following parameters: the location of the first orthogonal beam, the delay between the signals, and the distance between the beams. The locations of the orthogonal beams have been measured with a microscope which was fixed to a microscrew driven translator. The precision of the measured distance was 5 μm.

2.2.3. Mueller matrix of the SFC

In this section, the optical transfer function of the measurement system, i.e., laser, particle, cuvette, and photo-detector is described using the Mueller-matrix presentation.

The intensity and polarizing properties of the field are described via a Stokes vector (Collett (1993)). The transformation of the Stokes vector of incident light by interaction with a particle within the optical system can be described by a corresponding Mueller matrix. Hence, the complete information about the scattering from a particle can be presented in the form of a 4 × 4 Mueller matrix (Bohren and Huffman (1983)).

For the spherical particle the Mueller scattering matrix M_{sphe} is as follows (van de Hulst (1957)):

$$M_{\text{sphe}} = \begin{pmatrix} S_{11} & S_{12} & 0 & 0 \\ S_{12} & S_{11} & 0 & 0 \\ 0 & 0 & S_{33} & S_{34} \\ 0 & 0 & -S_{34} & S_{33} \end{pmatrix}, \quad (2.3)$$

where S_{ij} depends on polar scattering angle θ. Integration of scattered light at azimuthal angles from 0° to 360° is an inherent feature of the described cuvette. Integration over azimuthal scattering angles from 0° to 360° results in a Mueller scattering matrix M_{sys} (equation (2.4)), that describes the transformation of randomly polarized light scattered from a particle in the optical scanning system:

$$M_{\text{sys}} = \frac{1}{2} \begin{pmatrix} 2S_{11} & 0 & 0 & 0 \\ 0 & S_{11} + S_{33} & 0 & 0 \\ 0 & 0 & S_{11} + S_{33} & 0 \\ 0 & 0 & 0 & 2S_{33} \end{pmatrix}. \quad (2.4)$$

The polarization properties of the scattered light, which is detected by the PMT, is described by the Stokes vector V_{det} resulting from the multiplication of the matrix M_{sys} (equation (2.4)) and the Stokes vector V_{linpol} (equation (2.5)), which corresponds to linearly polarized light. Thus, the light intensity detected by the PMT is V_{det} (equation (2.6))

$$V_{\text{linpol}} = (1,1,0,0)^\top, \qquad (2.5)$$

$$V_{\text{det}} = (S_{11}, (S_{11}+S_{33})/2, 0, 0)^\top. \qquad (2.6)$$

For spherical particles, $S_{11} = 0.5(|S_1|^2 + |S_2|^2)$ and $S_{33} = 0.5(S_2^* S_1 + S_1^* S_2)$, where S_1 and S_2 are scattering functions for perpendicular and parallel polarization of incident light in relation to the scattering plane, respectively, and the asterisk denotes the complex conjugate. Finally we have to multiply the V_{det} vector with the vector $(1,0,0,0)$ (photomultiplier) that gives the element $S_{11}(\theta)$. It means that the SFC measures the angular dependency of the element S_{11} for linear polarized incident light. The scattering signal can be numerically simulated by calculating S_1 and S_2 from Mie theory (Section 1.1).

The Mueller scattering matrix M of an arbitrary shaped particle is as follows (van de Hulst (1957)):

$$M = \begin{pmatrix} S_{11} & S_{12} & S_{13} & S_{14} \\ S_{21} & S_{22} & S_{23} & S_{24} \\ S_{31} & S_{32} & S_{33} & S_{34} \\ S_{41} & S_{42} & S_{43} & S_{44} \end{pmatrix}. \qquad (2.7)$$

In general the S_{ij} depends on polar θ and azimuthal φ scattering angles, $S_{ij} = S_{ij}(\theta, \varphi)$. Integration over azimuthal scattering angles from $0°$ to $360°$ by the SFC optical cuvette results in scattering intensity

$$I_s = \int_0^{2\pi} S_{11}(\theta, \varphi) \, d\varphi \qquad (2.8)$$

and

$$I_s = \int_0^{2\pi} [S_{11}(\theta, \varphi) + S_{12}(\theta, \varphi) \cos(2\varphi) + S_{13}(\theta, \varphi) \sin(2\varphi)] \, d\varphi \qquad (2.9)$$

for non-polarized and linear polarized incident light, respectively.

These scattering intensities measured with the SFC can be simulated with the T-matrix approach (Section 1.2) for particles with an axial symmetry and with the Discrete Dipole Approximation (Section 1.3) for arbitrary shaped particles.

Chapter 2. Flow cytometry in measurement of light scattering

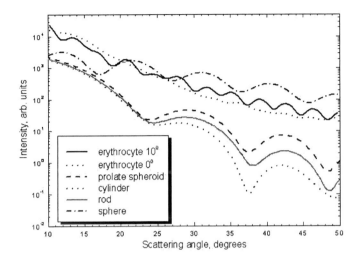

Figure 2.4. S_{11} element as a function of polar scattering angle of sphere, erythrocyte, cylinder, prolate spheroid, and rod calculated from Mie theory, T-matrix method, and Discrete Dipole Approximation. The light-scattering profiles of erythrocyte are shown in different orientation relating to direction of incident beam

In order to demonstrate polymorphism of LSPs caused by morphological characteristics of particles we calculated the angular dependency of S_{11} elements of Mueller matrix for a sphere, an erythrocyte, a cylinder, a prolate spheroid, and a rod. The results shown in Figure 2.4 demonstrate a sensitivity of S_{11} element of Mueller matrix to the particle shape and orientation that can be used in initial data to solve the ILS problem for individual particles.

2.3. POLARIZING SCANNING FLOW CYTOMETER

2.3.1. Optical set-up

The optical set-up of the SFC allows measurement of one LSP of individual particles. A structure of this angular function depends of particle morphology, orientation, and polarization of the incident beam. This fact makes available the solution of the ILS problem only for spherical particles or

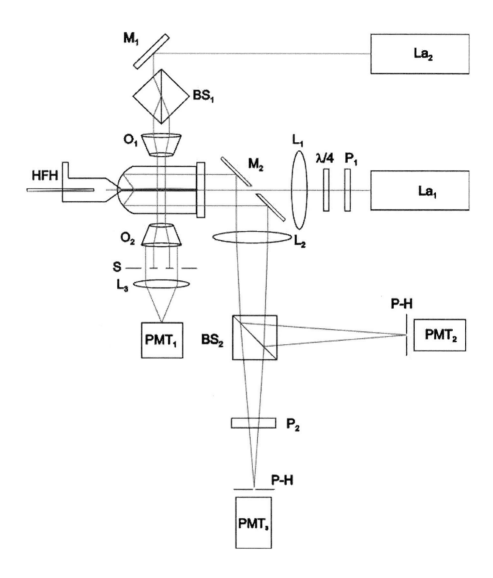

Figure 2.5. The schematic lay-out of the cuvette-assembly and optical system of the Polarizing Scanning Flow Cytometer. The following elements are shown: lasers (La), beam splitter (BS), lens (Le), hydrofocusing head (HFH), pin-hole (P-H), photomultiply tubes (PMT), mirror (M_1, M_2), mask (S), polarizers (P), and quart-wave plate ($\lambda/4$)

particles in fixed orientation. Indeed the solution of the ILS problem for prolate spheroid must provide determination of length, diameter, refractive index, and two orientation angles from one-dimensional function, for example $S_{11}(\theta)$ in Figure 2.4. Our theoretical simulation of light scattering of non-spherical particles showed that this problem could not be solved in sufficient precision for particle characteristics. Polarizing properties of scattered light should be utilized to provide a solution of the ILS problem by additional data. For this propose we developed Polarizing Scanning Flow Cytometer (PSFC) that allows one to measure at least two combinations of elements of Mueller matrix (equation (2.7)).

The schematic lay-out of the cuvette-assembly and optical system of the PSFC is shown in Figure 2.5. There are five new optical elements in PSFC comparing SFC (Figure 2.1) that are as follows: polarizers P_1 and P_2, quart-wave plate $\lambda/4$, beam splitter BS_2, and photomultiplier tube PMT_3. These elements give the additional light-scattering signal that can be used to solve the ILS problem for a particle with complex shape.

2.3.2. Mueller matrix

The complete information about the scattering from a particle can be presented in the form of a 4 × 4 Mueller-matrix (Collett (1993), Bohren and Huffman (1983)). The Mueller scattering matrix M of the arbitrary shaped particle is as follows (van de Hulst (1957)):

$$M = \begin{pmatrix} S_{11} & S_{12} & S_{13} & S_{14} \\ S_{21} & S_{22} & S_{23} & S_{24} \\ S_{31} & S_{32} & S_{33} & S_{34} \\ S_{41} & S_{42} & S_{43} & S_{44} \end{pmatrix}. \tag{2.10}$$

The optical set-up of the PSFC was shown in Figure 2.5. Each element of the PSFC set-up can be characterized with the Mueller matrix. The direction of the polarizer P_2 in Figure 2.5 determines the basic coordinate system. The Mueller matrix of the polarizer P_2 that sets in $0°$-orientation is as follows:

$$M_0 = \begin{pmatrix} 1 & 1 & 0 & 0 \\ 1 & 1 & 0 & 0 \\ 0 & 0 & 0 & 0 \\ 0 & 0 & 0 & 0 \end{pmatrix}. \tag{2.11}$$

The Mueller matrix of the polarizer P_1 that sets in 45°-orientation is as follows:

$$M_{45} = \begin{pmatrix} 1 & 0 & 1 & 0 \\ 0 & 0 & 0 & 0 \\ 1 & 0 & 1 & 0 \\ 0 & 0 & 0 & 0 \end{pmatrix}. \tag{2.12}$$

The Mueller matrix of the quart-wave plate is as follows

$$M_{qp} = \begin{pmatrix} 1 & 0 & 0 & 0 \\ 0 & 1 & 0 & 0 \\ 0 & 0 & 0 & -1 \\ 0 & 0 & 1 & 0 \end{pmatrix}. \tag{2.13}$$

The rotation of the basic coordinate system with angle φ is described by

$$M(\varphi) = \begin{pmatrix} 1 & 0 & 0 & 0 \\ 0 & \cos(2\varphi) & \sin(2\varphi) & 0 \\ 0 & -\sin(2\varphi) & \cos(2\varphi) & 0 \\ 0 & 0 & 0 & 1 \end{pmatrix}. \tag{2.14}$$

In order to evaluate the light scattering intensity on PMT_2 in Figure 2.5 we have to multiply the matrices $M(-\varphi) \times M \times M(\varphi) \times M_{qp} \times M_{45}$ with the vector $(1, Q, U, V)^\top$ (randomly polarized light) on the right side and with the vector $(1, 0, 0, 0)$ (photomultiplier) on the left side. Additionally the result of multiplication must be integrated over azimuthal angle φ because of spherical mirror of the optical scanning cuvette (Figure 2.5). The result of the multiplication and integration is as follows

$$I_{s2}(\theta) = \int (1, 0, 0, 0) \times M(-\varphi) \times M \times M(\varphi) \times M_{qp} \times M_{45} \times \begin{pmatrix} 1 \\ Q \\ U \\ V \end{pmatrix} d\varphi$$

$$= k \int_0^{2\pi} [S_{11}(\theta, \varphi) + S_{14}(\theta, \varphi)] \, d\varphi, \quad (2.15)$$

where k is the coefficient. It means that intensity measured by PMT_2 is proportional to the combination of elements of the scattering matrix (2.10).

Chapter 2. Flow cytometry in measurement of light scattering

For the second channel of the PSFC the intensity measured by the PMT$_3$ is resulted from multiplication of the matrices $M_{45} \times M(-\varphi) \times M \times M(\varphi) \times M_{qp} \times M_{45}$ and is proportional to the following combination of the elements of the scattering matrix (2.10):

$$I_{s3}(\theta) = \int (1,0,0,0) \times M_{45} \times M(-\varphi) \times M \times M(\varphi) \times M_{qp} \times M_{45} \times \begin{pmatrix} 1 \\ Q \\ U \\ V \end{pmatrix} d\varphi$$

$$= k_1 \int_0^{2\pi} [S_{11}(\theta,\varphi) + S_{14}(\theta,\varphi) + (S_{21}(\theta,\varphi) + S_{24}(\theta,\varphi)) \cos(2\varphi)$$
$$+ (S_{31}(\theta,\varphi) + S_{34}(\theta,\varphi)) \sin(2\varphi)] d\varphi, \quad (2.16)$$

where k_1 is the coefficient. By means of the electronic circuit of the PSFC the coefficients k (equation (2.15)) and k_1 (equation (2.16)) were equalized to k. The same electronic circuit subtracts $I_{s2}(\theta)$ and $I_{s3}(\theta)$ finalizing the PSFC output: ordinary SFC channel $-I_s(\theta) = I_{s2}(\theta)$ (equation (2.15)) and additional polarizing channel $I_{sp}(\theta)$

$$I_s(\theta) = \int_0^{2\pi} [S_{11}(\theta,\varphi) + S_{14}(\theta,\varphi)] d\varphi,$$

$$I_{sp}(\theta) = k \int_0^{2\pi} [(S_{21}(\theta,\varphi) + S_{24}(\theta,\varphi)) \cos(2\varphi) \quad (2.17)$$
$$+ (S_{31}(\theta,\varphi) + S_{34}(\theta,\varphi)) \sin(2\varphi)] d\varphi.$$

Finally the PSFC allows measurement two independent combination of Mueller matrix elements. The important performance of the PSFC relates to measurement of light scattering of spherical particles. The elements of Mueller matrix (2.10) do not depend of azimuthal angle φ and element $S_{14} = 0$ for a *spherical* particle that results to the following:

$$I_s(\theta) = 2\pi k S_{11}, \qquad I_{sp}(\theta) = 0, \quad (2.18)$$

that corresponds to the SFC. The equation (2.18) demonstrates a new performance of the PSFC — direct identification of non-spherical particles from light scattering. The signal I_{sp} must exceed the noise level for a non-spherical particle. We tested this feature of the PSFC experimentally with measurement of a mixture of spherical and non-spherical particles. The experiments are described in Section 4.2.

An alteration of the optical elements of the PSFC allows one to form another combination of Mueller matrix elements. In our experiments we remove the quart-wave plate from the optical set-up of the PSFC and substitute the output polarizer M_{45} with M_0. In order to take into account this change we have to remove the matrix M_{qp} (equation (2.13)) from multiplication. Consequently the measured signals satisfy the following equations: equation (2.15) becomes

$$I_{s2}(\theta) = k \int_0^{2\pi} [S_{11}(\theta, \varphi) + S_{12}(\theta, \varphi) \cos(2\theta) - S_{13}(\theta, \varphi) \sin(2\theta)] \, d\varphi \quad (2.19)$$

and equation (2.16) becomes

$$I_{s3}(\theta) = k_1 \int_0^{2\pi} [S_{11}(\theta, \varphi) - S_{21}(\theta, \varphi) \cos(2\varphi) + S_{31}(\theta, \varphi) \sin(2\varphi)$$
$$+ (S_{12}(\theta, \varphi) - S_{22}(\theta, \varphi) \cos(2\varphi) + S_{32}(\theta, \varphi) \sin(2\varphi)) \cos(2\varphi)$$
$$- (S_{13}(\theta, \varphi) - S_{23}(\theta, \varphi) \cos(2\varphi) + S_{33}(\theta, \varphi) \sin(2\varphi)) \sin(2\varphi)] \, d\varphi. \quad (2.20)$$

In order to simplify the analysis of the scattering measured with the current set-up of the PSFC we formed the combination of the measured signals that are as follows

$$I_s(\theta) = I_{s2}(\theta) = k \int_0^{2\pi} [S_{11}(\theta, \varphi) + S_{12}(\theta, \varphi) \cos(2\varphi)$$
$$- S_{13}(\theta, \varphi) \sin(2\varphi)] \, d\varphi, \quad (2.21)$$

$$I_{sp}(\theta) = \frac{I_{s3}(\theta)}{I_{s2}}.$$

The elements of Mueller matrix $S_{ij}(\theta, \varphi)$ do not depend of azimuthal angle φ and element $S_{22} = S_{11}$, $S_{13} = S_{31} = S_{32} = S_{23} = 0$ for *spherical* particles (equation (2.3)) that results to the following:

$$I_s(\theta) = 2\pi k S_{11}, \qquad I_{sp}(\theta) = 1 - S_{33}/S_{11}. \quad (2.22)$$

The second equations of the systems (2.18) and (2.22) can be used for verification of performance of the PSFC in measurement of polarizing properties of light scattered by individual particles. The testing experiments with measurement of element of S_{33} of spherical polymer beads are described in Section 4.2.

In order to demonstrate dependency of the polarizing signal $I_{sp}(\theta)$ caused by orientation of a particle we simulated the signal $I_{sp}(\theta)$ that is combination

Chapter 2. Flow cytometry in measurement of light scattering   33

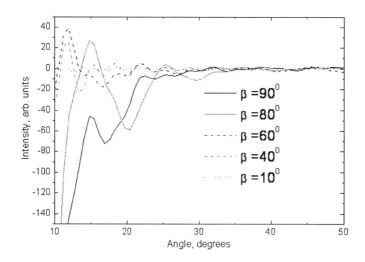

Figure 2.6. The scattering signal $I_{sp}(\theta)$ [equation (2.17)] simulated from discrete dipole approximation for single erythrocyte as a function of orientation angle β

of Mueller matrix elements of equation (2.17), with DDA for single mature erythrocyte. The orientation angle $\beta = 0$ corresponds to a coincidence of direction of the incident beam and symmetry axis of the erythrocyte. The results shown in Figure 2.6 demonstrate a sensitivity of signal $I_{sp}(\theta)$ to the particle orientation that can be used as an additional independent data to solve the ILS problem for individual non-spherical particles.

Thus the PSFC increases an amount of light-scattering data that can be used in a solution of the ILS problem for an individual particle with a complex shape and structure. Practically the PSFC set-up that measures the LSPs described by equation (2.17), allows direct identification of non-spherical particles from light scattering (equation (2.18)). A few approaches to solve the ILS problem using the experimental light-scattering data of SFC are presented in Chapter 3.

Chapter 3.

Inverse light-scattering problem of individual particles

The theoretical side of the Scanning Flow Cytometry is based on a solution of the inverse light-scattering (ILS) problem for individual particles. In particular parametric solutions of the ILS problem were developed for homogeneous spherical particles, absorbing spherical particles, spherical particles with centrally located nucleus, spheroidal particles. We advanced the neural network method for solution of the ILS problem for individual particles. The developed solution of the ILS problem provides a direct real-time determination of the physical characteristics of particles, individual particle characterization.

3.1. HOMOGENEOUS SPHERICAL PARTICLES

3.1.1. Nonlinear regression

Determination of dispersion parameters from light-scattering data requires an investigation of the complex inverse problem. The calculation of light-scattering intensity from spherical particles with a certain size and refractive index is rather complex, and therefore the exact expression of the inverse problem is generally not available in closed analytical form. The main method that was used by most authors to determine spherical particle parameters from light scattering is a fit of experimental light-scattering profile

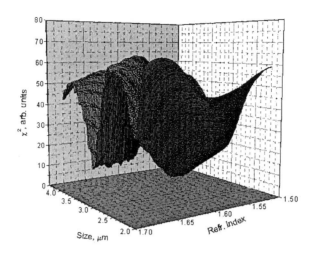

Figure 3.1. χ^2-surface of non-linear fit of experimental to theoretical light-scattering profiles of spherical particle

(LSP) to profiles calculated from Mie theory. This method requires significant computer time and a set of initial values of variable parameters of the fitting. Determination of parameters of nonspherical particles from the fitting is more complicated.

In our work we used non-linear fitting with the Levenberg–Marquardt method (Press et al. (1990)) to find the best fit of the theoretical Mie scattering pattern to the measured scattering pattern. The specific oscillations of the LSP of individual particles complicate the fit procedure dramatically. To demonstrate this fact we calculated χ^2 varying particle size, particle refractive index and normalising coefficient. The χ^2-surface as a function of particle size and refractive index is shown in Figure 3.1. There are a number of local minima on this plot. In order to exclude local minima from the fit solution we developed the logically dependent 3-pass algorithm. This algorithm allowed us to find the global minimum with the fit procedure. The initial parameters, size and refractive index, for non-linear fitting were retrieved from the parametric solution of the ILS problem described in the next section.

3.1.2. Parameterisation

The requirement in flow cytometry of a high analytical rate directed us to develop a method for particle analysis which should provide a real-time determination of particle characteristics and should be independent of instrument alignment. In this way, we tried to parameterise the ILS problem where spherical particle characteristics are calculated from the LSP parameters with approximating equations. The parameterisation means creation of approximating equations that relate LSP parameters to particle characteristics. Our solution arose from an analysis of the evolution of LSP extrema as a consequence of variation of particle parameters such as the size parameter $\alpha = \pi d m_0 / \lambda$ and relative refractive index $m = m'/m_0$, where d is the particle diameter, m_0 is the refractive index of the medium, λ is the wavelength of incident light, and m' is the refractive index of the particle. To find the parametric solution of the ILS problem using the LSP of an individual particle, one must take the following steps:

— choose LSP parameters that have differing sensitivities to variations in spherical particle characteristics (size parameter and relative refractive index);

— derive equations that relate the chosen LSP parameters to the particle characteristics; and

— determine errors in calculations of the particle parameters when the approximating equations are applied.

We defined a LSP parameter called fringe pitch that is angular distance between the first and second minima occurring after the boundary angle θ_b, $\Delta_2(\theta_b)$. For example, the fringe pitch of the LSP shown in Figure 3.2 is calculated from the formula $\Delta_2(15) = L_{\min 2}(15) - L_{\min 1}(15)$. The boundary angle θ_b plays an important role in development of the solution of the ILS problem. Historically introduction of the boundary angle was caused by the operational angular interval of the Scanning Flow Cytometer. On the other hand the boundary angle excludes the diffraction phenomena (Figure 1.4) from consideration because the boundary angle exceeds the critical angle defined by equation (1.21). The fringe pitch is not sensitive to variation of the particle relative refractive index and the approximating equation that relates the fringe pitch to the particle size can be derived using the direct light-scattering problem for spherical homogeneous particles, Mie theory. In order to obtain this approximating equation, a nonlinear fitting procedure

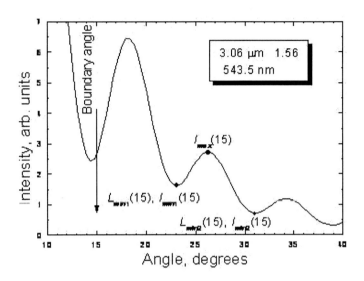

Figure 3.2. A light-scattering profile of the individual particle with a size of 3.06 μm and refractive index of 1.56. The wavelength and refractive index of medium used in the calculations are 543.5 nm and 1.333, respectively. The points marked by circle are used in the light-scattering profile parameter calculations

was applied to both the initial particle size and the size calculated from the tested approximating equations. A χ^2-test was used to minimize the residual standard error between initial and calculated particle sizes, and the final approximating equation corresponds to the minimum value of χ^2. The equation expressed in terms of the size parameter α is as follows:

$$\alpha = p_1 + p_2[\Delta_2(20)]^{-1} + p_3[\Delta_2(20)]^{-3}, \qquad (3.1)$$

where $\Delta_2(20)$ is measured in degrees, $p_1 = -1.05$ [0.11 s.d. (standard deviation)], $p_2 = 199.4$ (0.8 s.d.), $p_3 = 15$ (5 s.d.). Equation (3.1) yields a residual standard error of 0.66 for particle size parameters ranging from 5 to 80 and relative refractive indices ranging from 1.012 to 1.24. The coefficient p_2 gave a maximum dependency on α. This fact can be used to produce a simple approximating equation that relates the size parameter to the fringe pitch. We calculated LSPs from the Mie theory for size parameter α ranging from 4 to 100 with a step of 0.6 and for relative refractive index m ranging from 1.028 to 1.238 with a step of 0.008. The fringe pitch for each LSP has been determined and results of the calculation are shown in Figure 3.3. Each

Chapter 3. Inverse light-scattering problem of individual particles

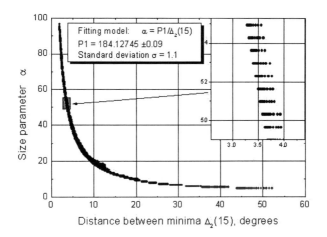

Figure 3.3. The size parameter as a function of the distance between minima of the light-scattering profile

point in this figure corresponds to the fixed pair of α and m. The nonlinear fitting procedure was applied to these data and the following approximating equation was recovered:

$$\alpha = p_2[\Delta_2(15)]^{-1}, \qquad (3.2)$$

where $p_2 = 184.13$ (0.09 s.d.). Equation (3.2) expressed in terms of particle size d is transformed to the following:

$$d = 1.023\lambda[\Delta_2(15)]^{-1} \approx \lambda/\Delta_2(15), \qquad (3.3)$$

where λ is the wavelength of incident light in medium and $\Delta_2(15)$ is the fringe pitch measured in radians. Equation (3.3) gives an accuracy of approximately $\lambda/2$ for size calculation. Equations (3.1)–(3.3) satisfy the requirements for real-time analysis of individual particles from light scattering with the scanning flow cytometry. As the fringe pitch is independent of an absolute intensity of scattered light, a calibration of the instrument that allows the measurement of the LSP of individual particle is not needed.

The next LSP parameter that can be used in the parametric solution must be more sensitive to the relative refractive index than to the size parameter. We chose forward visibility as a second parameter for determination of the refractive index of individual particles that is

$$(3.4) V_f(\theta_b) = (I_{\max} - I_{\min 1})/(I_{\max} + I_{\min 1}), \qquad (3.4)$$

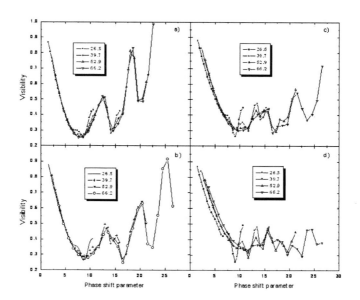

Figure 3.4. Forward visibility as a function of phase-shift parameter calculated for the different size parameters and for the (a) fourth, (b) fifth, (c) sixth, and (d) seventh pairs of minima and maxima

where $I_{\min 1}$ is the scattering intensity for the first minimum that occurs after the boundary angle θ_b and $I_{\max}$ is the scattering intensity for the next maximum (see Figure 3.2).

Unfortunately, the dependency of LSP visibility on relative refractive index is rather complex and does not provide good precision for determination of the refractive index of individual particles. To improve the precision of refractive-index determination, we analyzed the dependency of forward visibility on the phase-shift parameter of the particle defined as $\rho = 2\alpha(m-1)$. We calculated forward visibilities for four different size parameters and for different fixed pairs of LSP minima and maxima. These functions are shown in Figure 3.4. We can see from these figures that LSP forward visibility depends basically on the phase-shift parameter and is insensitive to the size parameter. The presented results allow us to claim that the parametric solution of the ILS problem can be carried out using LSP fringe pitch $\Delta_2(15)$ and forward visibility $V_f(15)$, which are relative to the particle size parameter α and phase-shift parameter ρ.

We calculated LSPs from the Mie theory for size parameter α ranging from 5 to 95 with a step of 0.6 and for phase-shift parameter ρ ranging from

Chapter 3. Inverse light-scattering problem of individual particles

0.5 to 16 with a step of 0.5. Analysis of these data forces us to introduce a new LSP parameter called an inverse visibility. The inverse visibility is calculated from the following formula:

$$V_i(\theta_b) = (I_{\max} - I_{\min 2})/(I_{\max} + I_{\min 2}), \qquad (3.5)$$

where $I_{\min 2}$ is the scattering intensity for the second minimum that occurs after the boundary angle θ_b and $I_{\max}$ is the scattering intensity for the previous maximum (see Figure 3.2). Both visibilities for each LSP have been determined from the LSPs mentioned above. Consequently, three LSP points used in the parametric solution of the ILS problem are calculated from the extremum triad marked by the circles in Figure 3.2. The visibilities for third and sixth triads as functions of phase-shift parameter are shown in Figure 3.5. These functions discover important phenomena: (1) both $V_f(15)$ and $V_i(15)$ are sensitive only to variations in phase-shift parameter for the fixed extremum triad; (2) the $V_f(15)$ and $V_i(15)$ give a different dependency on the phase-shift parameter ρ. In order to use these feature of the visibilities in the parametric solution of the inverse light-scattering problem we introduced the other LSP parameter that is the location of the first minimum that occurs after the boundary angle θ_b, $L(\theta_b)$. The $L(15)$ corresponds to the $L_{\min 1}(15)$ in Figure 3.2. Finally, the following four parameters, fringe pitch $\Delta_2(15)$, forward and inverse visibilities $V_f(15)$ and $V_i(15)$, and minimum location $L(15)$, allow the calculation of the spherical particle size and refractive index.

The size and phase-shift parameters of a spherical particle can be determined from the following algorithm:

(1) calculate the LSP parameters $\Delta_2(15)$, $V_f(15)$, $V_i(15)$, and $L(15)$;
(2) evaluate the parameter M from the following empirical equation:

$$M = \frac{L(15)}{\Delta_2(15)[V_i(15)]^{0.35}}, \qquad (3.6)$$

(3) using $V_f(15)$, $V_i(15)$, M and Table 3.1 calculate the size parameter α from one of two approximating equations that are as follows:

$$\alpha = p_1 + \frac{p_2[1 + p_3[V_i(15)]]}{\Delta_2(15)} + \frac{p_4}{[\Delta_2(15)]^3} + p_5[V_i(15)]^4, \qquad (3.7)$$

$$\alpha = p_1 + \frac{p_2[1 + p_3[V_f(15)]]}{\Delta_2(15)} + \frac{p_4}{[\Delta_2(15)]^3} + p_5[V_f(15)]^4, \qquad (3.8)$$

where p_t are the coefficients shown in Table 3.1.

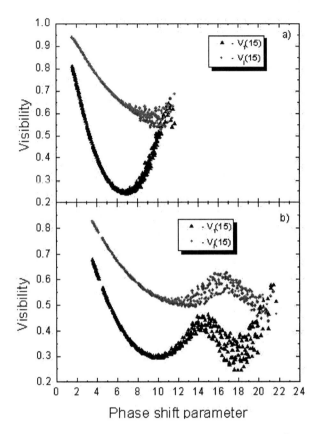

Figure 3.5. The forward and inverse visibilities as a function of phase-shift parameter for (a) third and (b) sixth triads of extrema

(4) using $V_f(15)$, $V_i(15)$ and M calculated above and Table 3.1, calculate the phase-shift parameter ρ from one of three approximating equations that are as follows:

$$\rho = p_1[V_i(15)] + p_2 + p_3[\Delta_2(15)][V_i(15)], \qquad (3.9)$$

$$\rho = p_1[1 + p_2[\Delta_2(15)]][1 - p_3\sqrt{V_f(15)}\,]\cos^{-1}\sqrt{\frac{V_f(15) - p_4}{1 - p_4}}, \qquad (3.10)$$

$$\rho = p_1[1 + p_2[\Delta_2(15)]][1 - p_3\sqrt{V_f(15)}\,]\left\{2\pi - \cos^{-1}\frac{V_f(15) - p_4}{1 - p_4}\right\}, \qquad (3.11)$$

where p_t are the coefficients shown in Table 3.1.

Table 3.1. The parameters of approximating equations used in calculation of particle size and phase-shift parameters

$0.7 < M < 1.44$; $V_f(15) < 0.183$	$0.7 < M < 1.44$; $V_f(15) > 0.183$; $V_i(15) > 0.802$	$0.7 < M < 1.44$; $V_f(15) > 0.183$; $V_i(15) < 0.802$
Eq. (3.7) for α with $p_1 = 0$; $p_2 = 248.13409$; $p_3 = -0.28936$; $p_4 = 0$; $p_5 = -0.41193$ Eq. (3.9) for ρ with $p_1 = -16.64153$; $p_2 = 18.77168$; $p_3 = -0.08651$	Eq. (3.7) for α with $p_1 = 0.05969$; $p_2 = 179.82336$; $p_3 = -0.0904$; $p_4 = 1094.4761$; $p_5 = 1.24749$ Eq. (3.10) for ρ with $p_1 = 2.97413$; $p_2 = -0.0078$; $p_3 = 0.42$; $p_4 = 0.1834$	Eq. (3.8) for α with $p_1 = 0$; $p_2 = 192.01097$; $p_3 = 0.08757$; $p_4 = 0$; $p_5 = 0$ Eq. (3.11) for ρ with $p_1 = 0.935$; $p_2 = 0.00151$; $p_3 = -0.21828$; $p_4 = 0.20408$
$1.44 < M < 2.52$; $V_f(15) < 0.238$	$1.44 < M < 2.52$; $V_f(15) > 0.238$; $V_i(15) > 0.716$	$1.44 < M < 2.52$; $V_f(15) > 0.238$; $V_i(15) < 0.716$
Eq. (3.7) for α with $p_1 = 0$; $p_2 = 245.75081$; $p_3 = -0.35735$; $p_4 = 87.26589$; $p_5 = 0$ Eq. (3.9) for ρ with $p_1 = -17.92801$; $p_2 = 18.47204$; $p_3 = -0.04219$	Eq. (3.7) for α with $p_1 = 0.1781$; $p_2 = 180.21491$; $p_3 = -0.00287$; $p_4 = 0$; $p_5 = 0.55652$ Eq. (3.10) for ρ with $p_1 = 4.06775$; $p_2 = -0.00364$; $p_3 = 0.43069$; $p_4 = 0.23106$	Eq. (3.8) for α with $p_1 = -0.16872$; $p_2 = 181.70602$; $p_3 = 0.23377$; $p_4 = 0$; $p_5 = 0$ Eq. (3.11) for ρ with $p_1 = 1.09628$; $p_2 = 0.00302$; $p_3 = -0.4648$; $p_4 = 0.24026$
$2.52 < M < 3.75$; $V_f(15) < 0.263$	$2.52 < M < 3.75$; $V_f(15) > 0.263$; $V_i(15) > 0.670$	$2.52 < M < 3.75$; $V_f(15) > 0.263$; $V_i(15) < 0.670$
Eq. (3.7) for α with $p_1 = 0$; $p_2 = 228.43185$; $p_3 = -0.30037$; $p_4 = 0$; $p_5 = -0.32932$ Eq. (3.9) for ρ with $p_1 = -26.69525$; $p_2 = 23.4812$; $p_3 = 0.08736$	Eq. (3.7) for α with $p_1 = 0$; $p_2 = 194.41051$; $p_3 = -0.1404$; $p_4 = 78.44302$; $p_5 = 2.51727$ Eq. (3.10) for ρ with $p_1 = 4.87332$; $p_2 = 0.00443$; $p_3 = 0.4321$; $p_4 = 0.26404$	Eq. (3.8) for α with $p_1 = 1.18537$; $p_2 = 156.71428$; $p_3 = 0.4164$; $p_4 = 247.11888$; $p_5 = -4.67776$ Eq. (3.11) for ρ with $p_1 = 1.12796$; $p_2 = 0.01529$; $p_3 = -0.69405$; $p_4 = 0.26389$

$3.75 < M < 5.05$; $V_f(15) < 0.300$	$3.75 < M < 5.05$; $V_f(15) > 0.300$; $V_i(15) > 0.650$	$3.75 < M < 5.05$; $V_f(15) > 0.300$; $V_i(15) < 0.650$
Eq. (3.7) for α with $p_1 = 0$; $p_2 = 225.68355$; $p_3 = -0.31772$; $p_4 = 0$; $p_5 = 1.91717$ Eq. (3.9) for ρ with $p_1 = -32.89565$; $p_2 = 26.44572$; $p_3 = 0.37281$	Eq. (3.7) for α with $p_1 = 0.34725$; $p_2 = 188.24396$; $p_3 = -0.10383$; $p_4 = 38.73535$; $p_5 = 2.56349$ Eq. (3.10) for ρ with $p_1 = 5.54932$; $p_2 = 0.01105$; $p_3 = 0.42642$; $p_4 = 0.29313$	Eq. (3.8) for α with $p_1 = 0$; $p_2 = 164.93857$; $p_3 = 0.47309$; $p_4 = 0$; $p_5 = -15.44728$; 0.31 s.d. Eq. (3.11) for ρ with $p_1 = 1.05073$; $p_2 = 0.0032$; $p_3 = -1.55917$; $p_4 - 0.3$
$5.05 < M < 6.30$; $V_f(15) < 0.310$	$5.05 < M < 6.30$; $V_f(15) > 0.310$; $V_i(15) > 0.600$	$5.05 < M < 6.30$; $V_f(15) > 0.310$; $V_i(15) < 0.600$
Eq. (3.7) for α with $p_1 = 0$; $p_2 = 268.67905$; $p_3 = -0.75304$; $p_4 = 100.42633$; $p_5 = 58.49569$ Eq. (3.9) for ρ with $p_1 = -40.40763$; $p_2 = 29.82812$; $p_3 = 0.90398$	Eq. (3.7) for α with $p_1 = 2.22086$; $p_2 = 178.71018$; $p_3 = -0.09579$; $p_4 = 47.41639$; $p_5 = 2.85551$ Eq. (3.10) for ρ with $p_1 = 6.28664$; $p_2 = 0.02438$; $p_3 = 0.4472$; $p_4 = 0.30345$	Eq. (3.8) for α with $p_1 = 0$; $p_2 = 158.99857$; $p_3 = 0.58985$; $p_4 = 0$; $p_5 = -38.16392$ Eq. (3.11) for ρ with $p_1 = 0.85174$; $p_2 = -0.00165$; $p_3 = -2.89996$; $p_4 = 0.31584$
$6.30 < M < 7.50$; $V_f(15) < 0.325$	$6.30 < M < 7.50$; $V_f(15) > 0.325$; $V_i(15) > 0.580$	$6.30 < M < 7.50$; $V_f(15) > 0.325$; $V_i(15) < 0.580$
Eq. (3.7) for α with $p_1 = 0$; $p_2 = 266.2522$; $p_3 = -0.76249$; $p_4 = 60.98683$; $p_5 = 78.97906$ Eq. (3.9) for ρ with $p_1 = -50.31642$; $p_2 = 34.38578$; $p_3 = 1.65343$	Eq. (3.7) for α with $p_1 = 2.23877$; $p_2 = 179.24101$; $p_3 = -0.08299$; $p_4 = 26.9403$; $p_5 = 3.06907$ Eq. (3.10) for ρ with $p_1 = 6.87893$; $p_2 = 0.04293$; $p_3 = 0.46257$; $p_4 = 0.31379$	Eq. (3.8) for α with $p_1 = 0$; $p_2 = 155.5301$; $p_3 = 0.65503$; $p_4 = 0$; $p_5 = -60.69687$ Eq. (3.11) for ρ with $p_1 = 0.65463$; $p_2 = -0.00474$; $p_3 = -4.90284$; $p_4 = 0.32492$

$7.50 < M < 8.81$; $V_f(15) < 0.340$	$7.50 < M < 8.81$; $V_f(15) > 0.340$; $V_i(15) > 0.567$	$7.50 < M < 8.81$; $V_f(15) > 0.340$; $V_i(15) < 0.567$
Eq. (3.7) for α with $p_1 = 7.12583$; $p_2 = 262.73187$; $p_3 = -1.04842$; $p_4 = 112.24863$; $p_5 = 154.74753$; 0.20 s.d. Eq. (3.9) for ρ with $p_1 = -59.65005$; $p_2 = 38.16305$; $p_3 = 2.94889$	Eq. (3.7) for α with $p_1 = 0$; $p_2 = 186.49335$; $p_3 = -0.06208$; $p_4 = 2.75872$; $p_5 = 2.90648$ Eq. (3.10) for ρ with $p_1 = 7.42432$; $p_2 = 0.05891$; $p_3 = 0.47032$; $p_4 = 0.32529$	Eq. (3.8) for α with $p_1 = 3.25955$; $p_2 = 129.56832$; $p_3 = 1.23819$; $p_4 = 0$; $p_5 = -176.09163$ Eq. (3.11) for ρ with $p_1 = 0.53578$; $p_2 = 0.00341$; $p_3 = -6.87507$; $p_4 = 0.34174$
$8.81 < M < 10.15$; $V_f(15) < 0.340$	$8.81 < M < 10.15$; $V_f(15) > 0.340$; $V_i(15) > 0.540$	$8.81 < M < 10.15$; $V_f(15) > 0.340$; $V_i(15) < 0.540$
Eq. (3.7) for α with $p_1 = 0$; $p_2 = 271.18396$; $p_3 = -0.86003$; $p_4 = 36.37673$; $p_5 = 157.78965$ Eq. (3.9) for ρ with $p_1 = -68.12487$; $p_2 = 41.67202$; $p_3 = 4.20451$	Eq. (3.7) for α with $p_1 = 0$; $p_2 = 188.16684$; $p_3 = -0.08026$; $p_4 = 2.13939$; $p_5 = 4.7196$ Eq. (3.10) for ρ with $p_1 = 7.83948$; $p_2 = 0.10193$; $p_3 = 0.49928$; $p_4 = 0.32924$	Eq. (3.8) for α with $p_1 = 0$; $p_2 = 153.53323$; $p_3 = 0.67826$; $p_4 = 0$; $p_5 = -107.8367$ Eq. (3.11) for ρ with $p_1 = 0.48535$; $p_2 = -0.0958$; $p_3 = -11.05834$; $p_4 = 0.33959$

Equations (3.7)–(3.11) are presented in an analytical form and allow the real time determination of size and refractive index of individual spherical particles. As the LSP parameters $\Delta_2(15)$, $V_f(15)$, $V_i(15)$, and $L(15)$ are independent of an incident light intensity, the above mentioned parametric solution permits a direct determination of the particle size and refractive index from light-scattering measurements. At present the parameterisation covers a significant region of the size versus refractive index map where the particle size and refractive index can be calculated from the light-scattering profiles (Figure 3.6).

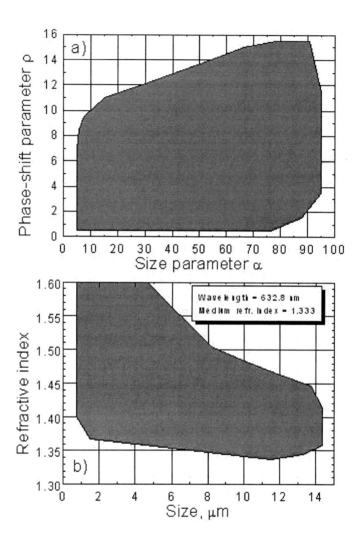

Figure 3.6. The region of size and phase-shift parameters map (a) where these parameters can be determined with the parametric solution of inverse light-scattering problem. The region of the size and refractive index map (b) was determined from the map (a) with the wavelength of 632.8 nm, He-Ne laser, and refractive index of medium of 1.333, water

Chapter 3. Inverse light-scattering problem of individual particles   47

3.1.3. Parameterisation from spectral decomposition

A parametric solution of the inverse light-scattering problem provides calibration-free and real-time characterization of individual spherical particles from light scattering. Unfortunately, this method is very critical to the quality of the experimental data, i. e. it gives significant errors processing noisy or distorted signals.

Traditionally noisy experimental signals are processed by means of a spectral approach that assumes the formation of a Fourier spectrum of experimental signals. The spectral approach to solve the inverse light scattering problem for individual homogeneous sphere was introduced by Ludlow and Everitt (1995). They used an expansion of the light-scattering profile (LSP) in the angular region ranging from 0° to 180° using a series of Gegenbauer functions. The cut-off point of the Gegenbauer spectrum was found to correlate with the size of the sphere by a unique dependence. The Fourier transform was used by Min and Gomez (1996) to determine sizes of droplets with known index of refraction. They measured the LSP with a photodiode array in angles from 9° to 18°. The array signal was processed with the fast Fourier transform that generates two values, frequency and phase, which correspond to the number and angular positions of the scattering lobes. These two values provide an accurate indicator of the particle size. Godefroy and Adjouadi (2000) proposed the flow imaging experimental setup where light scattering from individual particle was recorded by CCD camera into a 2-D light scattering pattern. They used 2-D fast Fourier transform to estimate the size of spherical particles.

From a practical point of view we are interested in developing a noise-insensitive method for particle sizing which can be realized with the SFC. The SFC allows measurement of the entire LSPs of individual particles within angular region ranging from 5 to 100 degrees. The LSPs of homogeneous spherical particles were calculated from Mie theory. The LSP of a particle with size parameter α of 53 and relative refractive index m of 1.16 is shown in Figure 3.7 a). The FFT was applied to the calculated LSP that gives the amplitude spectrum shown in Figure 3.7b). The location of the steep shoulder in the spectrum can be related to the particle size parameter. The analysis turns out to be problematic in the determination of the location of this steep shoulder for small particles. In order to solve this problem we multiply the LSP with a modifying function $F(\theta)$ shown in Figure 3.7a) by the dashed line. The resulting modified LSP is presented in Figure 3.7c). Finally, the corresponding FFT spectrum of the modified LSP is presented in Figure 3.7d). There are two clear advantages in this

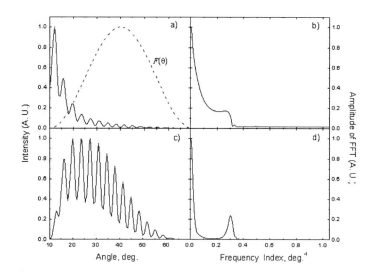

Figure 3.7. Light-scattering profile (a) with spectrum (b) of individual spherical particle. The dashed line represents the modifying function $F(\theta)$. The light-scattering profile modified by multiplication with the function $F(\theta)$ (c) and spectrum (d)

modified spectrum: 1) the steep shoulder is substituted with a clear peak; 2) the width of low-frequency part is reduced a few times.

It should be noted that some flexibility is available in the definition of the function $F(\theta)$. The requirements are as follows: 1) $F(\theta)$ should be continuous and monotonous; 2) $F(\theta_l) = 0$ where θ_l is the lowest angle of the LSP; 3) $\Delta\theta \leq W \leq \Theta$, where W is the width of $F(\theta)$, $\Delta\theta$ is the distance between LSP minima, Θ is the angular range of the LSP. In this work, we have defined the function $F(\theta)$ in the following form:

$$F(\theta) = \sin^2\left(\pi \frac{\theta - \theta_l}{\theta_h - \theta_l}\right), \qquad (3.12)$$

where θ_h is the highest angle of the LSP. In order to develop an algorithm that allows sizing of individual particles from the FFT spectrum of the modified LSP we have calculated LSPs for particles with size parameter ranging from 8 to 180 with a step of 1 (corresponding to size variation from 1.2 to 27.2 μm at the wavelength λ of 0.633 μm). The phase-shift parameter ρ was varied from 3 to 101 with a step of 2.8. Relative refractive indices of polystyrene beads and all biological cells are within the resulting

Chapter 3. Inverse light-scattering problem of individual particles

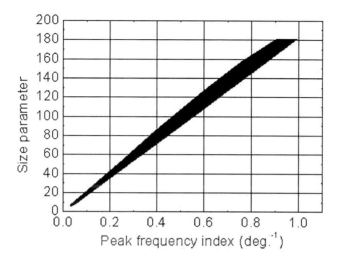

Figure 3.8. Size parameter of particles with different refractive indices $m = 1.39$, 1.49, and 1.70 as a function of peak frequency index P_f

range of refractive indices. The relation between the size parameter and the location of the peak in the frequency domain, P_F, for different refractive indices is shown in Figure 3.15. The slope of the line is increased with increasing refractive index. The unit of P_f is in reciprocal degrees. A linear equation can be used to describe the dependence of the size parameter α on P_F:

$$\alpha = kP_F, \tag{3.13}$$

where the coefficient $k = 191.32 \pm 0.04$ was obtained from linear regression. Equation (3.13) provides a standard deviation σ_α of 4.7 in size parameter determination that results in a mean relative error of 3.6 % for particle sizing from the FFT spectrum of the modified LSP.

The magnitude of the relative error is caused by a variation of refractive index of particles within the above mentioned bounds. In order to improve the precision of size determination we have to take into account the effect of refractive index on the algorithm developed. We considered the following situations: a) arbitrary refractive index; b) conditionally definable refractive index.

<u>Effect of refractive index: arbitrary refractive index.</u> With unknown refractive index of a particle we have to choose additional parameters of the FFT spectrum that are sensitive to variation of the refractive index. The

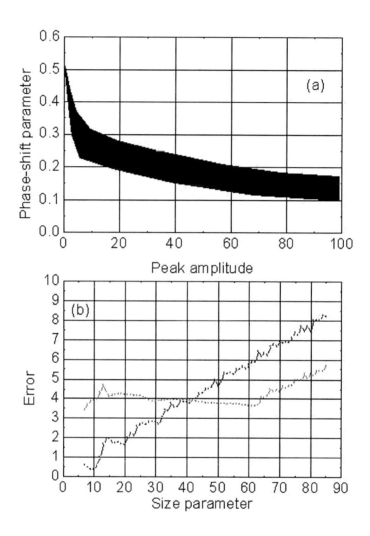

Figure 3.9. The scatter-plot of the peak amplitude as a function of size and phase-shift parameter (a). Maximal errors in calculation of size parameter from equation (3.13) (grey) and from equation (3.14) (black) (b)

peak amplitude A_F of the peak in the normalized FFT spectrum has been chosen. This parameter has to reduce the effect of refractive index on particle sizing from FFT spectrum. The scatter-plot of the peak amplitude as a function of phase-shift parameter for different sizes is shown in Figure 3.9a. This scatter-plot allowed us to choose the peak amplitude sensitive term in the approximating equation. The approximating equation, that relates the size parameter and FFT spectrum parameters, P_F and A_F, is as follows:

$$\alpha = k(1 + k_1/A_F)P_F + k_2, \qquad (3.14)$$

where the coefficients $k = 163.79 \pm 0.09$, $k_1 = 0.02334 \pm 0.00009$, $k_2 = 4.62 \pm 0.04$ were obtained from non-linear regression. Equation (3.14) provides a standard deviation σ_α of 2.3 in size parameter determination from the FFT spectrum of the LSP. The size of a particle can be determined from the proposed algorithm with a precision of approximately 0.35 μm. The additional FFT parameter has allowed us to increase the precision of sizing 2 times in comparison with equation (3.13). However the effect of refractive index is negligible for small size parameters. The maximum systematic errors in calculation of the size parameter from equations (3.13) and (3.14) as a function of the size parameter are shown in Figure 3.9b by grey and black lines, respectively. The following statement resulted from the functions: the effect of refractive index should be takes into account when a size parameter of analyzed particle exceeds 40. If size parameter less than 40, equation (3.13) can be successfully applied for particle sizing with spectral decomposition method. The size parameter of 40 corresponds to a 6 μm sphere suspended in water (refractive index of 1.333) and illuminated by red light (wavelength of 632.8 nm).

The spectral decomposition method is insensitive to experimental noise that results in more precise sizing of individual spherical particles from light scattering. Additionally this method reduces a negative effect of the boundary angle on precision of sizing. The spectral decomposition demonstrated a practical significance because at present two experimental SFCs are supported with this method for calibration-free real-time sizing. The spectral approach has a potential in determination of refractive index of spherical particle because of sensitivity of the phase-shift parameter to the peak amplitude of FFT spectrum.

3.1.4. Sub-micron spherical particles

The parametric solution introduced in Sections 3.1.2 and 3.1.3 permits the determination of the size and refractive index of spherically modelled particles over the ranges 1–15 μm and 1.37–1.6, respectively. The angular locations of two LSP minima are used in the parametric solution. A particle with a size of 1 μm produces the LSP, an incident radiation wavelength of 0.6328 μm, with angular locations of the first and second minima at approximately 38° and 67°, respectively. The ratio of scattering intensities $I(67°)/I(15°)$ is approximately 10^{-4}. From a practical point of view, measurement of the LSP of this particle results in considerable errors in the determination of the angular locations. Our experience in the determination of the particle size and refractive index forced us to modify the parametric solution for micron and sub-micron particles.

In this section, we describe the approach to determine individual submicron particle size and refractive index from a parametric solution of the inverse light-scattering problem. This parametric solution is based on approximating equations that relate the particle characteristics to the parameters of the LSP. These equations allow the real-time calculation of particle size and refractive index ranging from 0.5 to 1.5 μm and from 1.37 to 1.69, respectively.

3.1.4.1. Light-scattering profile parameters

A particle is characterized by a size parameter $\alpha = \pi d m_0/\lambda$ and phase-shift parameter $\rho = 2\alpha(m-1)$, where $m = m'/m_0$ is the relative refractive index, d is the particle diameter, m_0 is the refractive index of the medium, λ is the wavelength of incident light and m' is the refractive index of the particle. In order to choose LSP parameters that are appropriate to the parametric solution, we analyzed the formation of LSP structure with variation of these particle parameters. To calculate the LSPs we used a computer program that is based on the algorithm of Bohren and Huffman (1983). The exact solution of Mie scattering theory was used to calculate the LSP for non-absorbing spheres and non-polarized incident light. LSPs were calculated for the following parameters: size parameter α varying from 3.3 to 10 in steps of 0.3 and phase-shift parameter ρ varying from 0.2 to 5.32 in steps of 0.02. The analysis of the calculated LSPs showed that the size parameter depends mainly on the angular location $L(15)$, whereas the phase-shift parameter depends on the visibility

$$U(15) = \frac{I(15) - I_{\min}(15)}{I(15) + I_{\min}(15)}, \qquad (3.15)$$

where $I_{\min}(15)$ is the light-scattering intensity for the LSP minimum $L(15)$ and $I(15)$ is the light scattering intensity for the boundary angle 15° (Figure 3.10). The size parameter α as a function of the angular location $L(15)$

Chapter 3. Inverse light-scattering problem of individual particles   53

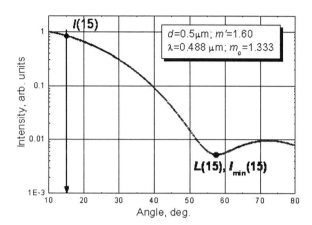

Figure 3.10. The light-scattering profile of individual spherical particles calculated from the Mie theory. The points marked by a circle were used in evaluation of the light-scattering profile parameters

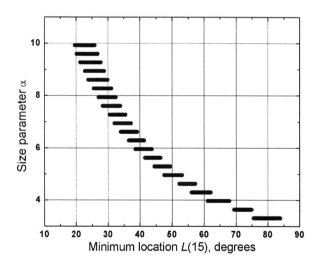

Figure 3.11. The size parameter as a function of angular location of the minimum $L(15)$. Each point on the graph corresponds to a fixed pair of size and phase-shift parameters

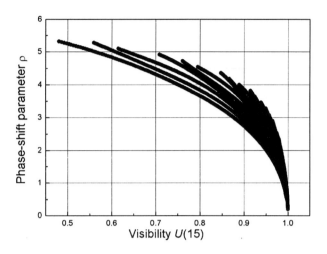

Figure 3.12. The phase-shift parameter as a function of visibility $U(15)$. Each point on the graph corresponds to a fixed pair of size and phase-shift parameters

and phase-shift parameter ρ as a function of the visibility $U(15)$ are shown in Figure 3.11 and 3.12, respectively. Each point on these figures corresponds to a fixed pair of the particle parameters α and ρ.

3.1.4.2. Approximating equations

In order to obtain the approximating equation that relates the LSP parameters, $L(15)$ and $U(15)$, to the particle size parameter, α, a non-linear regression was applied to both the initial size parameters and the size parameters calculated from the fitted equations. A χ^2 test was used to minimize the residual standard error between the initial and calculated particle size parameters. The following particular functional form for the approximating equation was chosen:

$$\alpha = \frac{p_1(1 + p_2 U(15) + p_3[U(15)]^3)}{L(15)} + p_4 + p_5[U(15)]^3 + \frac{p_6}{[L(15)]^3}, \quad (3.16)$$

where p_i are a set of fitting parameters. This choice was based on the analysis of the relevance of the LSP parameters $L(15)$ and $U(15)$ in the different power terms in equation (3.16). The fitting parameters p_i (Table 3.2) of the presented terms are most significant for the residual standard error in calculation of the particle parameter. The size parameter α is determined

Table 3.2. The fitting parameters of approximating equations for calculation of the size and phase shift parameters

Size parameter equation	Phase shift parameter equation
$\alpha = p_1(1 + p_2 U(15) + p_3[U(15)]^3)/L(15) + p_4 + p_5[U(15)]^3 + p_6/[L(15)]^3$	$\rho = 0.05[1 + q_1 L(15)^2 + q_2 L(15)^3 + q_3 L(15)^4][1 + q_4\sqrt{U(15)}] \times a\cos(\sqrt{U(15)})$
$p_1 = 216.20144$, err. $= 1.8$;	$q_1 = 0.5352$, err. $= 0.004$;
$p_2 = 1.22468$, err. $= 0.017$;	$q_2 = -0.01201$, err. $= 0.00008$;
$p_3 = -1.13964$, err. $= 0.015$,	$q_3 = 0.00007$, err. $= 5 \cdot 10^{-7}$,
$\sigma = 0.05$;	$\sigma = 0.06$;
$p_4 = -8.46102$, err. $= 0.09$;	$q_4 = 0.0566$, err. $= 0.009$
$p_5 = 8.90546$, err. $= 0.09$;	
$p_6 = 7222.25645$, err. $= 300$	

from equation (3.16) with a standard deviation σ of 0.05 (Table 3.2) with a in the range 3–10.

As the LSP visibility $U(15)$ basically depends on the phase-shift parameter, we produced the equation with a set of coefficients that relate the LSP parameters to ρ. The same non-linear regression was applied both to the phase-shift parameters that were evaluated for the LSP set and to the phase-shift parameters calculated from the fitted equations. A χ^2 test was used to minimize the residual standard error of the phase-shift parameters. The main equation for the phase-shift parameter calculation is

$$\rho = 0.05[1 + q_1 L(15)^2 + q_2 L(15)^3 + q_3 L(15)^4]$$
$$\times [1 + q_4\sqrt{U(15)}]a\cos(\sqrt{U(15)}), \quad (3.17)$$

where q_i are a set of fitting parameters. These parameters q_i are shown in Table 3.2. The phase-shift parameter ρ is determined from equation (3.17) with a standard deviation σ of 0.08 (Table 3.2) with ρ in the range 0.2–5.3.

Equations (3.16) and (3.17) were used in the determination of particle size and refractive index from the LSPs measured with the SFC.

3.2. SPHERICAL PARTICLES WITH ABSORPTION

3.2.1. Light-scattering profile parameters

The parameterisation can be successfully applied to solve the ILS problem for spherical particles with absorption of particle matter at the wavelength of the incident light. We developed this approach in analysis of erythrocytes from light scattering. A chemically sphered erythrocyte can be modelled by a homogeneous sphere that absorbs the incident light, 488 nm, by haemoglobin.

To calculate the intensities and angular locations of the scattering intensity local extrema, we have used a computer program that is based on the algorithm of Bohren and Huffman (1983). The exact solution of the Mie-scattering theory was used to calculate the LSP for absorbing spheres for unpolarized incident light. In our calculations the particle size d was varied from 4 to 7.5 μm with a step of 0.1 μm that corresponds to size parameters α from 26 to 50. In calculations, the hemoglobin concentration of the RBC's HbC was also varied from 5 to 45 g/dl with a step of 0.5 g/dl that corresponds to phase-shift parameters $\rho = 2\alpha(m_R - 1) = 2\alpha\beta \times HbC$ from 0.4 to 6.3 and to absorption parameters $\varepsilon = 2\alpha m_I = d\sigma \times HbC \cdot 10^{-4}$ from $1.5 \cdot 10^{-5}$ to $2.5 \cdot 10^{-4}$, where $m_R = m'_R/m_0$, $m_I = m'_I/m_0$.

The LSPs were calculated for the wavelength of the incident light, $\lambda = 632.8$ nm, and the refractive index of the medium, $m_0 = 1.333$. The phase-shift parameter of a particle means a change in a phase between waves passing through the particle substance with refractive index m_R and path length d and surrounding medium with refractive index m_0 and path length d. The absorption parameter means an attenuation of the amplitude of the wave, when the wave is passing through a particle substance with a refractive index m_R and a path length d.

The local minimum and the local maximum points of the LSP shown in Figure 3.13 are used in the LSP parameter calculations. The fringe pitch and the forward visibility are defined as $\Delta_2(15) = L_{\min 2} - L_{\min 1}$ and

$$V_f(15) = \frac{I_{\max} - I_{\min 1}}{I_{\max} + I_{\min 1}}, \qquad (3.18)$$

respectively, where $L_{\min 2}$ and L_{mml} are the angular locations of the second and the first minima behind a boundary angle of 15 deg, respectively. In Section 3.1 we showed that $\Delta_2(15)$ and $V_f(15)$ are the LSP parameters that give the minimum residual standard error for calculation of the particle characteristics. In the current study the fringe pitch $\Delta_2(15)$ and the forward visibility $V_f(15)$ are also used. For the region of size parameter mentioned above, the LSP minimum intersects the boundary angle. This results in different dependencies of the phase-shift parameter on forward visibility. Here we introduce an additional LSP parameter that is an angular location of the first minimum occurring after an angle of 15 deg $L(15)$. By plotting

Chapter 3. Inverse light-scattering problem of individual particles

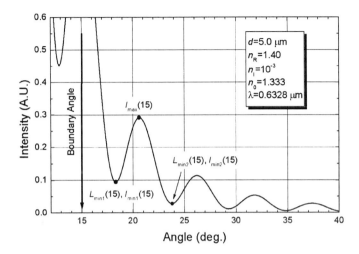

Figure 3.13. Light-scattering profile of a single homogeneous sphere calculated from Mie theory. The refractive index of the surrounding medium is 1.333, water

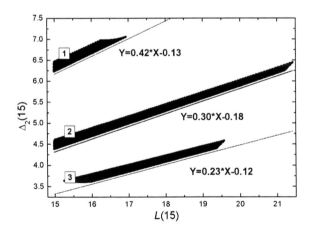

Figure 3.14. Parametric plot. The light-scattering profile parameters $\Delta_2(15)$ and $L(15)$ were evaluated from the profiles that were calculated within such a size and refractive-index region that corresponds to the characteristics of a typical erythrocyte. The plot region is divided into three subregions by lines, and the corresponding equations are shown on the plot

calculated pairs $\Delta_2(15)$ and $L(15)$ for calculated α and ρ, we can conclude that the region of the fringe pitch $\Delta_2(15)$ versus minimum location $L(15)$ map is broken into three parts. These regions of the $\Delta_2(15)$ and $L(15)$ map are shown in Figure 3.14. The dependency of the particle characteristic α on $\Delta_2(15)$ is shown in Figure 3.15. The dependency of the particle characteristic, $\rho = 2\alpha\beta \cdot HbC$ on $V_f(15)$ is shown in Figure 3.16. Each point on these figures corresponds to a pair of α and ρ. These data demonstrate that the LSP fringe pitch is not sensitive to variations of ρ, whereas the LSP visibility is not sensitive to variations of α. Whereas the size parameter α has a uniform dependency on the fringe pitch (Figure 3.15), the phase-shift parameter depends on the forward visibility in a different manner (Figure 3.16) for each subregion of the $\Delta_2(15)$ and $L(15)$ map.

3.2.2. Approximating equations

To obtain the approximating equation that relates the LSP parameters to the particle-size parameter α, a nonlinear fitting procedure was applied to both the initial size parameters and the size parameters calculated from the fitted equations. We fitted the dependencies in Figure 3.15 and Figure 3.16 by different particular functional forms to obtain the approximation equations for α and ρ, respectively. A χ^2 test has been used to minimize the residual standard error between the initial and the calculated particle-size parameters. The following particular functional form for the approximating equation has been chosen

$$\alpha = \frac{p_1[1 + p_2[V_f(15)]]}{\Delta_2(15)} + p_3[V_f(15)]^4, \qquad (3.19)$$

where p_i is a set of fitting parameters that are different in each region in Figure 3.13. These parameters are presented in Table 3.3. The choice of functional form was based on the analysis of the relevance of the LSP parameters $\Delta_2(15)$ and $V_f(15)$ in the different power terms in equation (3.19). The fitting parameters p_i of the presented terms are most significant for the residual standard error in the particle parameter calculations. Especially the parameter p_2 is determined with a minimum relative error (Table 3.3). The standard deviations σ_α of the size parameter determination from equation (3.19) are also shown in Table 3.3 for each region.

Because the LSP visibility depends mostly on the phase-shift parameter, we have produced an equation with a set of coefficients that relate the LSP parameters to ρ. The same nonlinear fitting procedure was applied to

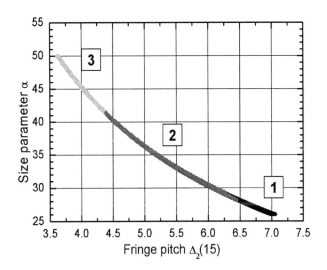

Figure 3.15. Size parameter as a function of the fringe pitch $\Delta_2(15)$. The points marked by 1, 2, and 3 correspond to the subregions in Figure 3.14

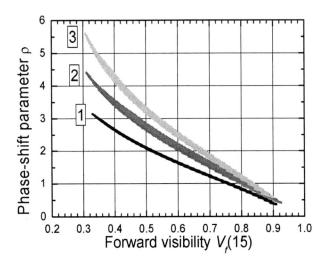

Figure 3.16. Phase-shift parameter as a function of the forward visibility $V_f(15)$

both, the phase-shift parameters that were evaluated for the LSP set and the phase-shift parameters calculated from the fitted equations. A χ^2 test has been used to minimize the residual standard error of the phase-shift parameters. The main equation for the phase-shift parameter calculation is as follows:

$$\rho = q_1[1 + q_2[\Delta_2(15)]][1 - q_3[V_f(15)]]\cos^{-1}\sqrt{\frac{V_f(15) - q_4}{1 - q_4}}, \qquad (3.20)$$

where q_i is a set of fitting parameters. The need for the $\cos^{-1}$ function is based on the fact that the dependency of ρ on the visibility is oscillating (Figure 3.5). The coefficients q_i are shown in Table 3.3. The standard deviations σ_ρ of the determination of the phase-shift parameter from equation (3.20) are also shown in Table 3.3 for each region. The hemoglobin concentration of the RBC is calculated from the determined size and phase-shift parameters by using the definition of the phase-shift parameter, $\rho = 2\alpha\beta \times HbC$.

The size and the hemoglobin concentration of an erythrocyte can be determined from the following algorithm:

1. Calculate the parameters $L(15)$, $\Delta_2(15)$ and $V_f(15)$) from the measured LSP.

2. Determine the number of the region from the parametric plot $\Delta_2(15)$ versus $L(15)$ (Figure 3.14).

3. Calculate the size parameter α and the phase-shift parameter ρ from equation (3.19), equation (3.20), and Table 3.3.

4. The size d and hemoglobin concentration HbC of the RBC are found from the definition of α and ρ.

We have developed a method for simultaneous measurement of size and concentration of absorbing matter of individual spherical particles. We have applied an algorithm that allows construction of approximating equations and that was previously developed for analysis of homogeneous spherical particles, to construct the equations relating the LSP parameters to the absorbing particle characteristics. The parametric equations have been developed for the measurement of sphered erythrocyte volume and hemoglobin concentration. However, the algorithm can be easily used to construct the approximating equations for other types of absorbing particles, although one has to take into account the coefficient β and the absorption parameter ε.

Chapter 3. Inverse light-scattering problem of individual particles 61

Table 3.3. The approximating equations
for calculation of the size and phase-shift parameters

Size parameter equation	Phase shift parameter equation
$\alpha = p_1[1+p_2[V_f(15)]]/\Delta_2(15) + p_3[V_f(15)]^4$	$\rho = q_1[1+q_2[\Delta_2(15)]][1-q_3[V_f(15)]]$ $\times \cos^{-1}\sqrt{(V_f(15)-q_4)/(1-q_4)}$
$p_1 = 181.11315$, 0.015 error	$q_1 = 2.38$, 0.07 error
$p_2 = 0.00231$, 0.00019 error	$q_2 = 0.042$, 0.004 error $\sigma_\rho = 0.018$
$\sigma_\alpha = 0.009$	$q_3 = 0.741$, 0.003 error
$p_3 = 0.4001$, 0.0044 error	$q_4 = 0$
$p_1 = 181.72445$, 0.015 error	$q_1 = 2.276$, 0.013 error
$p_2 = -0.01106$, 0.00019 error	$q_2 = 0.061$, 0.001 error $\sigma_\rho = 0.027$
$\sigma_\alpha = 0.025$	$q_3 = 0.7023$, 0.0013 error
$p_3 = 0.80578$, 0.0056 error	$q_4 = 0.251$, 0.002 error
$p_1 = 182.40653$, 0.019 error	$q_1 = 2.485$, 0.014 error
$p_2 = -0.0209$, 0.00024 error	$q_2 = 0.0942$, 0.0017 error $\sigma_\rho = 0.03$
$\sigma_\alpha = 0.04$	$q_3 = 0.719$, 0.001 error
$p_3 = 1.22948$, 0.011 error	$q_4 = 0.3000$, 0.0005 error

3.3. SPHERICAL PARTICLES WITH A COVER

A spherical particle with a cover, two concentric spheres, plays a significant role in modelling of biological cells because of cellular nucleus. In this section we apply the spectral decomposition approach (Section 3.1.3) to find a parametric solution of the ILS problem for a spherical particle with a cover.

The S_{11} elements of the Mueller matrix of spherical particles with a cover were calculated from the computer program that is based on the algorithm of Bohren and Huffman (1983). A particle is characterized by a diameter d, a refractive index of shell m_s, a diameter of core d_c, and a core refractive index m_c. We consider particles with dense core, $m_c > m_s$, modelling a cell with a nucleus. The LSPs were calculated within the regions of particle characteristics ranging from 43 to 86 for size parameter α ($\alpha = \pi d m_0/\lambda$, where λ is the wavelength of the incident light, m_0 is refractive index of medium), from 0.5α to 0.7α for a core size parameter α_c, and from 7 to 18 for an effective core refractive index ε, $\varepsilon = 2\sqrt{\alpha}\sqrt{\alpha}(m_c/m_0 - 1)$. The refractive index of a shell was fixed at 1.36 corresponding to the refractive index of cellular cytoplasm.

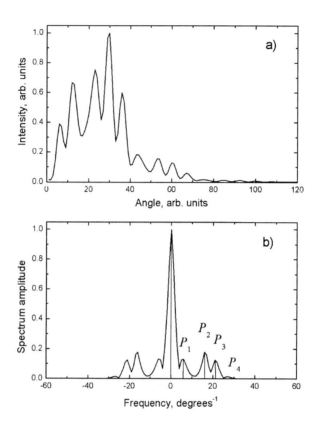

Figure 3.17. Modified light-scattering profile of a particle modeled by two concentric spheres ($\alpha = 53$, $m_s = 1.02$, $\alpha_c = 33$, $m = 1.14$ (a), spectrum amplitude of the modified light-scattering profile (b)

3.3.1. Light-scattering profile parameters

According to the spectral decomposition procedure an LSP was multiplied by the modifying function equation (3.12). The FFT was applied to the modified function to analyze a spectrum of a spherical particle with a cover. The typical modified LSP and corresponding Fourier spectrum of analyzed particle are shown in Figure 3.17. The spectrum was normalized at 1 for 0th frequency. There are four peaks in this spectrum instead of single peak comparing the spectrum of homogeneous sphere shown in Figure 3.7d). Analysis of migration of the peaks in spectrum of covered sphere LSP with varia-

Chapter 3. Inverse light-scattering problem of individual particles

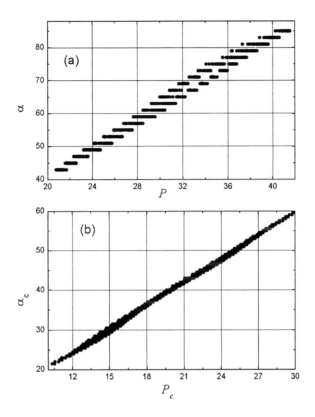

Figure 3.18. Size parameter α as a function of the parameter P of the light-scattering profile of covered sphere (a). Core size parameter α_c as a function of the P_c for ε ranging from 7 to 18 (b)

tion of particle characteristics allowed us to choose the following parameters for parametric solution of the ILS problem for two concentric spheres: $P = 2P_3 - P_2$, $P_c = P_2$, and $C = 1/A_3$, where P_i and A_i is the location and amplitude of i-peak, respectively. The particle characteristics as a function of the LSP parameters are shown in Figure 3.18 and Figure 3.19.

3.3.2. Approximating equations

e calculated LSPs from the exact theory for size parameter α ranging from 43 to 86 with a step of 1.0, for a core size parameter α_c ranging from 0.5α to

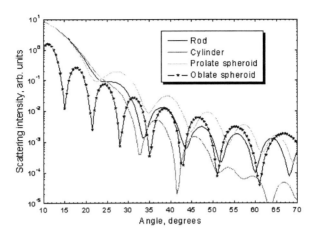

Figure 3.19. Particle characteristic ε as a function of parameter C for different α and α_c

0.7α with a step of 0.5, and for an effective core refractive index ε ranging from 7 to 18 with a step of 0.75. The parameters P, P_c, and C for each LSP have been determined and results of the calculation are shown in Figure 3.18 and Figure 3.19. Each point in these figures corresponds to the fixed triad α, α_c, and ε. The nonlinear fitting procedure was applied to these data and the following equations were recovered:

$$\alpha = p_1 C + p_2(1 + p_3 C)P, \qquad (3.21)$$

where $p_1 = -0.14901$, $p_2 = 2.21533$, $p_3 = -1.33551$ ($\sigma = 0.87$);

$$\alpha_c = q_1 C + q_2(1 + q_3 C)P_c, \qquad (3.22)$$

where $q_1 = -0.11134$, $q_2 = 2.04195$, $q_3 = 2.42502$ ($\sigma = 0.31$);

$$\varepsilon = s_1(1 + s_3 P_c) + s_2(1 + s_4 P_c)C, \qquad (3.23)$$

where $s_1 = 27.96529$, $s_2 = -40.44168$, $s_3 = -0.00503$, $s_4 = -0.01053$ ($\sigma = 0.7$).

In this section we introduced the parametric solution of the ILS problem for two concentric spheres. The covered sphere is a suitable model for nucleated biological cells (monocytes, lymphocytes). The equations (3.21)–(3.23) allow one to determine a size of the cell, a size of the nucleus, and

Chapter 3. Inverse light-scattering problem of individual particles   65

a density of the nucleus (nucleus refractive index). The solution is based on an LSP of cells measured in the angular interval ranging from 10 to 70 degrees. Experimentally this approach can be realized by means of scanning flow cytometry introduced in Chapter 2.

3.4. SPHEROIDAL PARTICLES

Spheroidal particles play an important role in modelling of biological particles. For instance a number of microorganisms can be modelled by a prolate spheroid whereas morphology of blood platelets satisfies geometry of oblate spheroids. In this section we have introduced the parametric solution of the ILS problem for prolate spheroids oriented along of direction of incident light. This orientation can be realized with scanning flow cytometer because of hydrodynamic focussing of a flow carrying particles analyzed.

3.4.1. Particle characteristics

In order to solve the ILS problem for oriented spheroids with parametric approach we studied a formation of the LSP under a variation of particle characteristics. The LSP were calculated from the T-matrix method (Section 1.2). According to our expertise in parameterisation a choice of particle characteristics affects on precision of parametric solution. We have defined three particle characteristics from physical properties of a prolate spheroid, length l, diameter d and relative refractive index m. A prolate spheroid was characterized by the equal-volume size parameter, $\alpha = 2\pi d/\lambda$, where d is the diameter of an equal-volume sphere and λ is the wavelength of incident laser beam, long-axis phase-shift parameter, $\rho = 2\pi l(m-1)/\lambda$, where l is the length of the long axis, and m is the relative effective refractive index of the particle, and the axis ratio, $\varepsilon = d/l$, where d is the particle diameter.

3.4.2. Light-scattering profile parameters

The following LSP parameters for a prolate spheroid were chosen to solve the ILS problem: location of the first LSP minimum occurring after the boundary angle 15°, $L(15)$; visibility, $V(\varphi) = (I_{max} - I_{min\,1})/(I_{max} + I_{min\,1})$, where $I_{min\,1}$ is the scattering intensity for the first minimum that occurs after the boundary angle 15° and I_{max} is the scattering intensity for the next maximum, and squeeze, $S(15) = (L_{min\,2}(15) - L_{min\,1}(15))/(L_{min\,1}(15))^{0.8}$, where $L_{min\,2}(15)$ is the scattering intensity for the second minimum that occurs after the boundary angle 15°.

LSPs of the prolate spheroids were calculated for the following characteristics: length l varying from 0.76 μm to 4.32 μm with a step of 0.032 μm, diameter d varying from 0.5 μm to 1.2 μm with a step of 0.07 μm, and refractive index m' varying from 1.35 to 1.45 with a step of 0.01. Taking into account the light wavelength of 0.6328 μm and water refractive index of 1.333, the defined particle characteristics cover the following regions: the equal-volume size parameter α from 4.9 to 15.8; long-axis phase-shift parameter ρ from 0.2 to 4.6; axis ratio ε from 0.3 to 0.9. The LSP parameters $L(15)$, $V(15)$, $S(15)$ were evaluated from the set of LSPs. The LSP parameters as functions of particle characteristics are shown in Figure 3.20. The results shown in Figure 3.20 demonstrate a reality of parametric solution of the ILS problem for prolate spheroids because of different dependency of the particle characteristics on the LSP parameters. For instance, the equal-volume size parameter α basically depends of location of the first LSP minimum $L(15)$ (Figure 3.20c) whereas the long-axis phase-shift parameter ρ depends of the visibility $V(15)$ (Figure 3.20b). This fact is in agreement with parametric solution of the ILS problem for sub-micron spherical particles (Section 3.1.4). The new dependency has been retrieved for a relation between the axis ratio ε and the LSP squeeze $S(15)$ (Figure 3.20a).

3.4.3. Approximating equations

Because location of the first LSP minimum, the LSP visibility, and LSP squeeze depend mostly on the equal-volume size parameter, the long-axis phase-shift parameter, and axis ratio, respectively, we have produced equations with a set of coefficients that relate the LSP parameters to particle characteristics. The nonlinear fitting procedure was applied to both, the particle characteristics that were evaluated for the LSP set and the particle characteristics calculated from the fitted equations. The nonlinear regression was applied to construct the approximating equations which are as follows:

$$\alpha = -p_1 + \frac{p_2}{L(15) + p_3 S(15)} + p_4 V(15)^2, \tag{3.24}$$

$$\rho = q_1(1 - q_2 L(15) - q_3 S(15))(1 - q_4 \sqrt{V(15)}) \cos^{-1} \sqrt{\frac{V(15) - q_5}{1 - q_5}}, \tag{3.25}$$

$$\varepsilon = t_1 + t_2(S(15) - t_3 L(15) + t_4 V(15)) \\ + t_5 \sqrt{S(15) - t_3 L(15) + t_4 V(15)}, \tag{3.26}$$

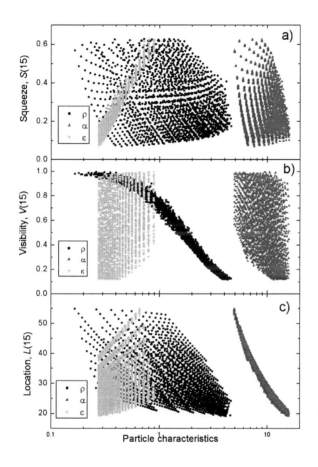

Figure 3.20. The light-scattering profile parameters (a) squeeze $S(15)$, (b) visibility $V(15)$, 1st minimum location $L(15)$ as function of particle characteristics equal-volume size parameter α, long-axis phase-shift parameter ρ, axis ratio ε

where $p_1 = 2.7167$, $p_2 = 369.83276$, $p_3 = 4.4657$, $p_4 = 0.95355$, $q_1 = 4.62095$, $q_2 = 0.00373$, $q_3 = 0.32788$, $q_4 = 0.62008$, $q_5 = 0.06402$, $t_1 = 0.22021$, $t_2 = 0.532$, $t_3 = 0.00161$, $t_4 = 0.0811$, $t_5 = 0.77528$.

The equations (3.24)–(3.26) give a good correlation (Figure 3.20) between the LSP parameters and the particles characteristics. The particle characteristics are determined from these equations with mean errors of 13 % for the particle length, 10 % for the particle diameter, and 1 % for the particle relative refractive index.

At present the parametric solution of the ILS problem for oriented prolate spheroids is unique method to determine the particle characteristics from light-scattering data measured experimentally. The non-linear regression utilized in characterization of spherical particles is not available because the T-matrix method requires a huge time for calculation of the LSP of single particle. The iteration process could not be realized at present with realistic duration. Fortunately our latest study of performance of neural networks in solution of the ILS problem gives a chance for development of a neural net that allows determination of non-spherical particle characteristics from light-scattering profiles. There is a subject of the next section.

3.5. NEURAL NETWORK

Solution of the inverse light-scattering problem for particulate matter plays a rising role with modern experimental technique that utilizes last achievements in optics technologies, electronics, and computers. At present technique for light-scattering measurements of particles gives an additional opportunity in a choice of methods for particle characterization, determination of particle characteristics from solution of the inverse light-scattering problem. In particular the Scanning Flow Cytometry described in Chapter 2 supports experimental data in a form which is sufficient and appropriate for use of computer oriented mathematical algorithms. The neural network is one of widespread methods in last decade.

The radial basis function (RBF) neural network was applied by Ulanowski et al (1998) to solve the inverse light-scattering (ILS) problem for spheres. They demonstrated the possibility of recovery of both the size and the refractive index of particles from angle-dependent light scattering data. The network was trained within intervals ranging from 0.5 μm to 1.5 μm and from 1.5 to 1.7 for size and refractive index, respectively. Size and refractive-index recovery was fast, taking typically 0.1 s on an average desktop computer. The work made by Wang et al (1999) demonstrated the ability of RBF networks to approximate high dimensional non-linear functions, commonly an ill-posed problem, and more importantly, it opened up an area associated with solving the ILS problem. RBF networks which employ clustering for locating hidden unit receptive field centres can achieve a performance comparable to back propagation networks.

Back-propagation network is probably the most widely used algorithm for generating classifiers and is often used for benchmarking other learning algorithms (Simon (1999)). A back-propagation neural network is a feed-

Chapter 3. Inverse light-scattering problem of individual particles 69

forward multi-layer neural network. The network has two stages: a forward pass and a backward pass. The forward pass involves presenting a sample input to the network and letting activations flow until they reach the output layer. The activation function can be any function. The linear sum, sigmoid function and Gaussian function are three often used functions. During the backward pass, the network's actual output (from the forward pass) is compared with the target output and error estimates are computed for the output units. The weights connected to the output units can be adjusted in order to reduce those errors. We can then use the error estimates of the output units to derive error estimates for the units in the hidden layers. Finally, errors are propagated back to the connections stemming from the input units. Detailed descriptions of the back-propagation algorithm can be found in Jain at al (2000) and Simon (1999).

The back propagation network is better to use when training data is expensive (or hard to generate) and/or retrieval speed, assuming a serial machine implementation, is critical (the smaller back propagation network size requires less storage and leads to faster retrievals compared to RBF networks). However, if the data is cheap and plentiful and if on-line training is required, then the RBF network is superior.

There were a few problems that relate to a fit of the neural network described by Wang et al (1999) to experimental technique. For instance, the input light-scattering data were normalized on scattering intensity in forward direction, at scattering angle of 0° that is outside of operation angular interval of the SFC. The second problem is arisen from requirements of flow cytometry in high analytical rate of analysis of individual particles, 10^3 particles per second. RBF networks use a large number of units in hidden layer that reduce the analytical rate. In order to solve the ILS problem we used back propagation network with threshold function $p(x) = x/(|x|+0.1)$ that was adapted to experimental traces of the Scanning Flow Cytometer.

The most important advantage of the neural network approach is an absence of strict analytical equations. On the other hand, training time and accuracy in determination of output parameters with a network depend on a choice of input and output parameters of the network. Our experience in analysis of formation of the LSP structure under a variation of particle characteristics has helped us to form the proper vectors of input and output parameters of the network. We have used the input parameters, parameters of the LSP, which are most insensitive to experimental noise, give a different dependency on variation of different particle characteristics, and are independent on absolute light-scattering intensity. The last requirement provides the calibration-free procedure of particle characterization from light scattering. Actually the neural network approach is similar to the parametric solution of the ILS problem removing a retrieval of approximating equations from the development procedure.

3.5.1. Spherical particles

The solution for spherical particles arose from an analysis of the evolution of LSP extrema as a consequence of variation of particle parameters such as the size parameter $\alpha = \pi d m_0 / \lambda$ and phase-shift parameter $\rho = 2\alpha(m-1)$ where $m = m'/m_0$, d is the particle diameter, m_0 is the refractive index of the medium, λ is the wavelength of incident light, and m' is the refractive index of the particle. The pair (α, ρ) forms the output vector of the network. The input vector of the network is formed from parameters of the normalized and modified LSPs. The LSP, S_{11} element of Mueller matrix, calculated from the Mie theory in the angular interval ranging from 10 degrees to 70 degrees, was normalized integrating over the angular interval. The angular interval corresponds to operational angular interval of the Scanning Flow Cytometer. The normalized LSP was modified by multiplication with function $F(\theta)$ defined in Section 3.1.3 (equation (3.12)). The resulted LSP was used to form the input parametric vector of the neural network that is as follows: locations of two maximal peaks of the LSP, location of the maximal peak of the LSP spectrum, slope of the LSP, first twenty points of the LSP.

We calculated LSPs from the Mie theory for size parameter α ranging from 8 to 100 with a step of 0.6 and for phase-shift parameter ρ ranging from 0.5 to 20 with a step of 0.5. The neural network was trained with a two-component noise: the noise of radiation source and the background noise. The input parameters of the network were evaluated from the LSPs pre-calculated. The back propagation network with the threshold function $p(x) = x/(|x| + 0.1)$ was formed by a 10 neuron hidden layer. This threshold function provides the best precision in determination of single spherical particle characteristics. The network has been trained with ten thousands LSPs generated randomly and covers the region on the phase-shift parameter versus size parameter map shown in Figure 3.21. Common PC trained the network in 5 hours.

The neural network was tested with ten thousands LSPs generated from pairs (α, ρ) randomly. This method yields a residual standard error of 1.9 for particle size parameters ranging from 8 to 100. The neural network approach has demonstrated a good precision in determination of size parameter from

Chapter 3. Inverse light-scattering problem of individual particles   71

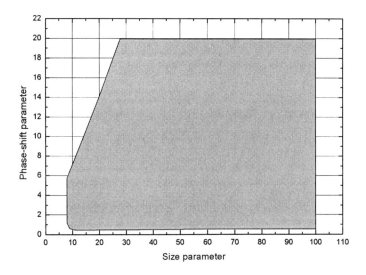

Figure 3.21. The phase-shift parameter versus size parameter map where the solution of the inverse light-scattering problem for single spherical particles with the neural network is available

LSPs. However the phase-shift parameter was determined with a residual standard error of 1.6 that is unsatisfactorily for particle characterization. In order to reduce the error we formed three overlapping sub-networks ranging in phase-shift parameter from 0.5 to 10, from 6 to 14, and from 10 to 20. This two-level network provides a residual standard error of 0.9. The doubled time of calculation is a price paid for the improvement.

The developed neural network covers substantial region of size versus refractive index map (Figure 3.22). The neural network approach allowed us to retrieve two parameters of individual spherical particles, size and refractive index. This performance of the neural network solution of the ILS problem conforms to the parametric solution. However the network method offers an advantage that is as follows: the operational region is not splitted by sub-regions similar to parametric solution (conditions from Table 3.1). These sub-regions respond in empty areas on the sub-region borders when wide-distributed spherical particles are measured. At present the SFC is supported by programming codes that include the neural network routine allowing a user to characterise spherical particles from experimentally measured LSPs in real-time regime excluding a calibration procedure from the

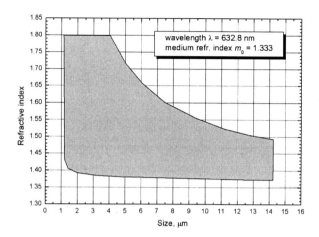

Figure 3.22. The region of refractive index versus size map where the neural network approach is available to solve the inverse light-scattering problem for individual spherical particles

instrument adjustment. The particles can be analyzed with a typical rate of 500 particles per second. We applied the developed neural network in analysis of spherical polymer beads and sphered red blood cells (Section 4.1).

Chapter 4.

Applications

First of all, the Scanning Flow Cytometer (SFC) can be used in the same applications where the ordinary flow cytometry performs an analysis of individual particles. The extended capacity of the SFC in the individual particle analysis can be demonstrated with the applications where the exact knowledge of particle parameters is required. The simulation of the ordinary flow cytometry can be easily performed with the SFC integrating the light-scattering profile (LSP) around 10° (forward signal) and 70° (side signal) of the scattering angle. In this section we consider a few applications where the SFC has been used.

4.1. POLYMER PARTICLES

At present, the polymer particle is used in many applications, generally in the field of medicine and biology (Frengen et al, 1995; Lindmo et al, 1990; Newman et al, 1992). Consequently, the polymer particle certification plays an important role for the particle production. The electron microscopy provides a sufficient accuracy in size measurements but it requires a long sample preparation time and powerful computers for image analysis. In order to demonstrate applicability of the SFC in the plastic bead certification we have used polystyrene particles (carboxylated polystyrene uniform latex microspheres, Duke Scientific, Cat. No. C300A). The size parameters (mean size of 3.06 μm and standard deviation of 0.18 μm) of these particles were specified by the manufacturer.

The LSPs of individual polystyrene particles were measured at polar angles from 10° to 80°. The LSP of one measured particle is shown in Fig-

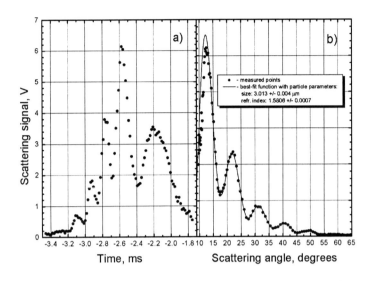

Figure 4.1. The light-scattering profile of one measured particle, (a) the experimental scattering-intensity record as a function of time, (b) the normalized scattering pattern using the transformation and normalization functions of the cuvette

ure 4.1. The experimental scattering-intensity trace as a function of time is presented in Figure 4.1a. Using the transformation and normalization functions as described in the Section 2.1.2, this trace was transformed into the normalized scattering pattern that is shown in Figure 4.1b (points). The numerical algorithm of Bohren and Huffman (1983) was used to derive the parameters S_1 and S_2, as described in the Section 2.1.3. Values of S_1 and S_2 were used to calculate the theoretical Mie scattering patterns of homogeneous spherical particles. Non-linear fitting using the Levenberg-Marquardt method was used to find the best fit of the theoretical Mie scattering pattern to the measured scattering pattern (Press et al, 1990). The initial parameters, size and refractive index, for non-linear fitting were retrieved from the parametric solution of the inverse light-scattering (ILS) problem.

Two hundred particles have been measured with the SFC and analyzed with the parameterisation. The results of this analysis are shown in Figure 4.2 and have given good agreement with the manufacturer specification for a bead size. We can conclude that the parameterisation gives a substantial simplification for bead certification compared with electron microscopy. Additionally, the parameterisation has given the refractive index distribution for these beads.

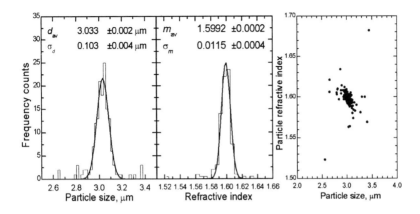

Figure 4.2. The polystyrene beads certified with the scanning flow cytometer. Distributions in size (a), in refractive index (b), refractive index vs size map (c)

The LSPs of polystyrene beads measured with the SFC were processed with the parametric solution (Section 3.1.2) and spectral decomposition method (Section 3.1.3). A typical LSP, modified LSP, and spectral decomposition of a polystyrene bead are shown in Figure 4.3. The parametric solution utilizes the distance between minima that occurs after the boundary angle of 15 degrees. The LSP presented in Figure 4.3a (open points) gives a distance between minima $\Delta\varphi_1(15)$ of 7.19 degrees. The spectral decomposition (Figure 4.3b) gives the frequency index of 0.145 degrees^{-1} that corresponds to a distance between minima $\Delta\varphi$ of 6.89 degrees. The distance between minima $\Delta\varphi$ is a superposition of distances between minima within the region ranging from 10 to 70 degrees that reduces an effect of the boundary angle on accuracy of individual particle sizing.

In order to demonstrate an effect of experimental noise on particle sizing of spectral decomposition we added 10% white noise to the indicatrix presented in Figure 4.3a. The parameterisation and spectral methods gave the $\Delta\varphi_1(15) = 7.34$ degrees and $\Delta\varphi = 6.87$ degrees, respectively. This means that the new approach is noise-insensitive and we have to expect improvement of accuracy of individual particle sizing with the spectral decomposition.

The size distributions for the samples of polystyrene microspheres and sphered erythrocytes were processed using the spectral approach and parametric solution of the ILS problem, and the results are presented in Fig-

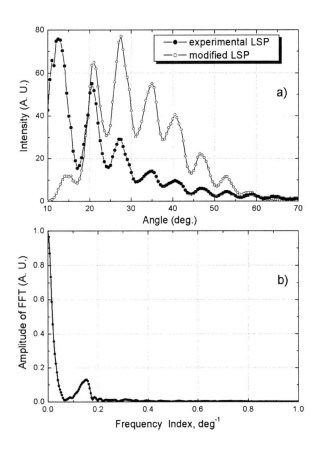

Figure 4.3. Typical light-scattering profile (a) (open points) of polystyrene bead measured with SFC. Modified light-scattering profile (a) (solid points) and spectral decomposition (b)

ure 4.4 and Table 4.1. The mean sizes retrieved from equation (3.13) are less than 6 μm, size parameter of 39.6 that gives an advantage of equation (3.13) over equation (3.14). This statement is in agreement with distribution parameters shown in Table 4.1. The distribution parameters retrieved from equation (3.13) are close to manufacturer specifications than the parameters retrieved from equation (3.14). However difference between mean sizes of sphered erythrocytes for both equations is negligible because the mean sizes are close to 6 μm where systematical errors are coincident. Indirect conformation of noise-insensitivity of the spectral method is resulted from smaller distribution widths retrieved from equation (3.13) than the widths obtained from parameterisation (Table 4.1).

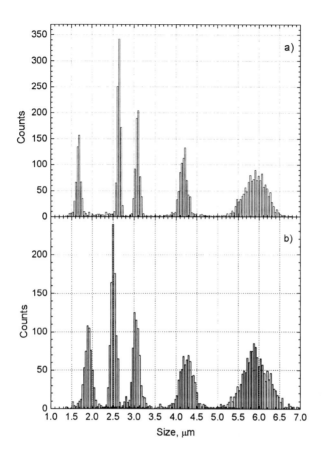

Figure 4.4. Size distributions of polystyrene latex microspheres and sphered erythrocytes, obtained in two different ways: (a) based on spectral decomposition and (b) based on parameterisation

The same measured LSPs were processed with neural network solution of the ILS problem (Section 3.5.1). The result of the analysis is shown in Figure 4.5. The particle parameters are retrieved from the LSP with a rate of 500 particles per second that provides a real-time analysis of individual spherical particles with the Scanning Flow Cytometer.

A specialised SFC can be equipped with specific neural network, for instance the SFC designed for characterization of erythrocytes with determination of volume and haemoglobin concentration of individual erythrocytes must be supported with the neural network trained on the region of param-

Table 4.1. Comparison of different equations
of spectral decomposition and parametric methods for particle sizing

N	Equation (3.13), mean diameter, μm	Equation (3.13), distribution width, μm	Equation (3.14), mean diameter, μm	Equation (3.14), distribution width, μm	Parameterisation, mean diameter, μm	Parameterisation, distribution width, μm
1	1.67	0.10	2.25	0.14	1.91	0.16
2	2.64	0.08	3.14	0.09	2.50	0.12
3	3.08	0.09	3.63	0.12	3.04	0.15
4	4.18	0.18	4.61	0.25	4.25	0.33
5	5.92	0.56	6.06	0.50	5.91	0.59

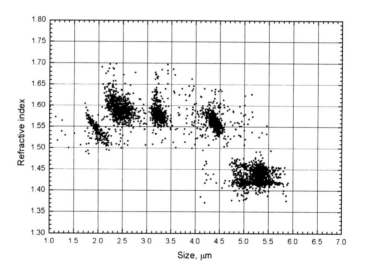

Figure 4.5. The refractive index versus size map of polymer particles and sphered erythrocytes measured with Scanning Flow Cytometer and processed with neural network solution of the inverse light-scattering problem

eters that corresponds to red blood cells. The SFC designed for certification of polymer beads must be supported with appropriate neural network for bead characterization. A specialized neural network will give best precision for individual particle characterization from light scattering.

4.1.1. Sub-micron polymer beads

The applicability of the parametric solution was demonstrated with analysis of polystyrene particles, the LSPs of which were measured with an SFC. We used an optical set-up and data acquisition system similar to that described in Section 2.1.

In order to verify our approximation equations, we produced polystyrene particles by dispersion polymerization (Chernyshev et al, 1997). We measured samples of two monodisperse polystyrene particles. The typical LSPs of these particles are shown in Figure 4.6a, b. The measured LSPs were processed with the parametric solution of the ILS problem.

The LSP parameters $L(15)$ and $U(15)$ (Section 3.1.4.1) were calculated from the LSPs. The size and phase-shift parameters of the particle were evaluated from equations (3.16) and (3.17), respectively. Taken into account the wavelength of the incident light, 488 nm, and medium refractive index, 1.333, the size and refractive index were determined from the definition of the particle parameters. The resulting sizes and refractive indices are shown in Figure 4.6. Additionally, these LSPs were processed with non-linear regression to the Mie theory. The best-fit LSPs are shown in Figure 4.6 (solid lines) with size and refractive index of the particles presented in the same figure. The disagreement between particle parameters obtained using these methods is caused by errors in the calculation of the LSP parameters $L(15)$ and $I_{\min}(15)$. These parameters were determined from a parabolic approximation of an LSP part whereas the best-fit parameters were retrieved from a non-linear approximation to the Mie theory for the full LSP. Moreover the particle parameters were determined from equations (3.16) and (3.17) with a systematic error. Hence the size and refractive index of particles in Figure 4.6 determined from the best-fit LSP are more realistic.

The size distribution of the polystyrene particles presented in Figure 4.7a was built by processing 300 particles with the parametric solution. We determined the mean size and standard deviation for both monodisperse particle samples. Consequently, polymer particles can be easily certified with the parametric solution of the inverse light-scattering problem. Additionally, the parametric solution allows us to visualize the result of the measurements

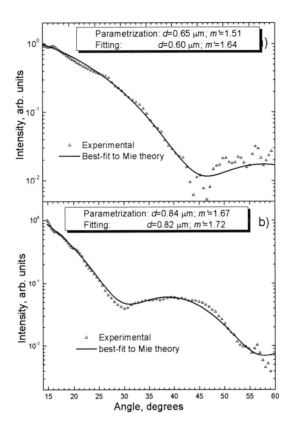

Figure 4.6. The typical experimental light-scattering profile (triangles) of individual polymer particles. The best-fit light-scattering profile (solid line) calculated from the Mie theory fits the experimental one. The size and refractive index of the particle calculated from the parametric solution and fitting procedure are presented

as a size versus refractive index map. This map is shown in Figure 4.7b. The distribution in refractive index for measured samples is caused by errors in the determination of the LSP parameters. Particles with a mean size of 0.66 μm give a scattering intensity half that given by particles with a mean size of 0.86 μm. This difference results in a larger signal-to-noise ratio for the large particles, especially in the region of the LSP minimum (Figure 4.6), and, hence, to a small distribution in refractive index.

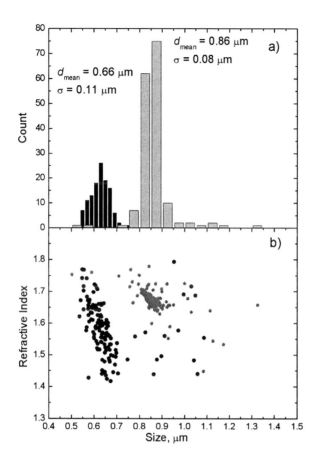

Figure 4.7. (a) The size distribution of two polystyrene particle samples measured with the scanning flow cytometer and processed with the parametric solution of the inverse light-scattering problem. (b) The size versus refractive index map for the same samples

4.2. POLYMER BEADS ANALYZED WITH POLARIZING SCANNING FLOW CYTOMETER

We tested the performance of the PSFC in simultaneous measurement of ordinary light-scattering profile (LSP) and polarizing LSP.

The PSFC was equipped with He–Ne laser for measurement of light scattering. The optical set-up of this PSFC corresponds to the optical set-up described in the Section 2.2.1. According to the theory of the PSFC the polarizing signal should be zero for spherical particles (equation (2.18)). In

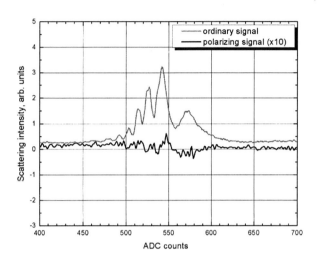

Figure 4.8. Ordinary and polarizing traces of a polystyrene bead measured with PSFC. The polarizing trace has been multiplied by 10

order to test this feature of the PSFC we measured the certified polystyrene beads. By means of the electronic circuit we set the polarizing signal close to zero. The typical ordinary and polarizing traces measured with the PSFC are shown in Figure 4.8. The ordinary trace corresponds to the trace of spherical particles measured with the SFC whereas the polarizing trace looks like fluctuations around zero. On the other hand the inidentified particle found in the same sample produced the ordinary and polarizing traces shown in Figure 4.9. The ordinary trace of that particle does not correspond to a spherical particle because of variable distance between extremes. At the same time the amplitude of the polarizing trace exceeds the amplitude of the polarizing trace of the spherical particle 10 times. This fact is in agreement with a prediction of light scattering theory — polarizing intensity is very sensitive to orientation of non-spherical particles. This experiment demonstrated the performance of the PSFC in selection of non-spherical particles from measurement of polarizing light scattering.

In another experiment, the PSFC was equipped with a He-Cd laser for measurement of light scattering. The optical set-up of this PSFC does not contain the quarter wave plate and, hence, output signals are described by equations (2.22). According to these equations the polarizing signal for

Chapter 4. Applications

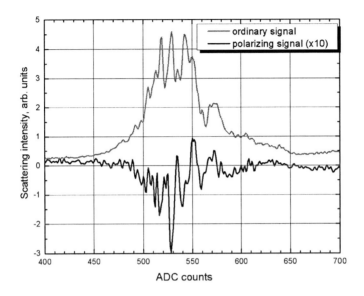

Figure 4.9. Ordinary and polarizing traces of an inidentified particle measured with PSFC. The polarizing trace has been multiplied by 10

spherical particles corresponds to the following combination of Mueller matrix elements:

$$I_{sp}(\theta) = 1 - S_{33}/S_{11}. \tag{4.1}$$

The value of expression (4.1) was calculated from the Mie theory for a particle with a size and refractive index of 3 μm and 1.6, respectively. The result of computing simulation of light scattering is shown in Figure 4.10. We would like to emphasize two peculiarities of the polarizing LSP: a) the polarizing signal tends to zero for forward direction and becomes sizeable for side light scattering; b) the oscillations for ordinary and polarizing LSPs are asynchronous. In order to test the performance of the PSFC in measurement of polarizing properties of light scattered by individual particles we measured the certified polystyrene beads with a mean size of 3 μm. The measured ordinary and polarizing LSPs are presented in Figure 4.11.

The typical experimental ordinary and polarizing traces (Figure 4.11a) of the individual bead were transformed into ordinary and polarizing LSPs (Figure 4.11b). Comparing the curves in Figure 4.10 and Figure 4.11b we are able to conclude that experimental and theoretical LSPs are in a good

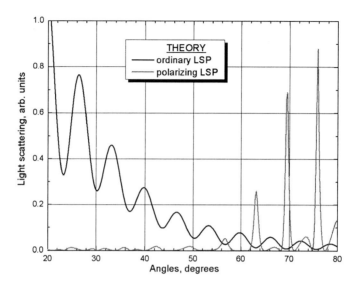

Figure 4.10. The ordinary and polarizing light-scattering profiles of individual spherical particle calculated from the Mie theory

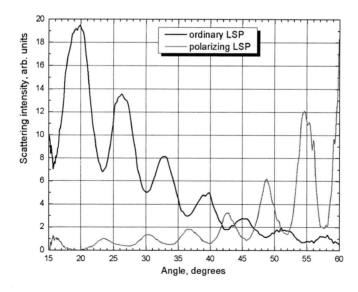

Figure 4.11. Experimental ordinary and polarizing light-scattering profiles measured with PSFC

agreement for the both ordinary and polarizing LSPs. This experiment demonstrated the performance of the PSFC in precise measurement of polarizing light scattering from individual particles. Analysis of polarizing LSPs would open new ways in solution of the inverse light-scattering problem to determine particle characteristics from light scattering.

4.3. POLYMERIZATION

The SFC was successfully applied to the study of dispersion radical polymerization. The developed mathematical model of the dispersion radical polymerization (Chernyshev et al, 1997) allows us to produce the analytical formula that relates the total volume of the growing particles and reaction time. One thousand particles that form the experimental size distribution were measured for every sample taken from the reaction volume at the fixed time. In order to demonstrate the applicability of the mathematical model, the evolution of the size distribution of growing polystyrene particles was processed with the parameterisation. The size distributions for the four time points are shown in Figure 4.12. The total volume of latex particles was evaluated as a function of time, then these data were fitted by the analytical formula and kinetic parameters of the polymerization were retrieved. The parameterisation provides an effective and an express analysis of the samples.

4.4. MILK FAT PARTICLES

The parameterisation was used for analysis of the fat content in natural milk. As usual, the size of milk fat particles is distributed between 1 and 10 μm with a maximum of about at 3 μm. The size of protein molecules ranges from 0.01 to 0.2 μm that is under the threshold of the trigger channel of SFC. Concentration of bacterial cells is more than three orders of magnitude lower than in the concentration of fat particles. A sample of fresh raw milk was diluted by a factor of $1.1 \cdot 10^4$ with water to provide suitable concentration of fat particles for the analysis with SFC. The internal flow rate was $1.003 \cdot 10^4$ ml/s. Approximately 1000 particles were measured for each sample. The value of the size and the mass was calculated for each particle on line as well as the total weight of the particles in each sample. The measurements were repeated for the same samples seven times and the results of the measurements are shown in Table 4.2. The fat content was 3.93 % with standard deviation of 0.34 % (absolute) when 1000 particles were

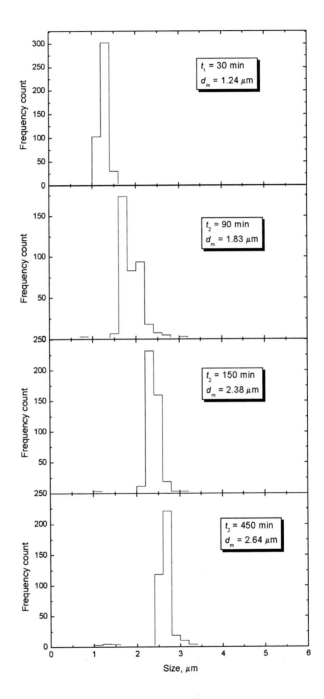

Figure 4.12. The size distributions for dispersion polymerization of styrene in isopropanol measured with the SFC and processed with parameterisation

Table 4.2. Analysis of fat contents in milk
with the scanning flow cytometer

Sample no	Sample weight, g	Total weight of fat particles, g	Fat content, %
1	1.056E-6	3.795E-8	3.59
2	1.061E-6	3.751E-8	3.53
3	1.045E-6	4.152E-8	3.97
4	1.049E-6	3.842E-8	3.63
5	1.051E-6	4.433E-8	4.22
6	1.059E-6	4.429E-8	4.18
7	1.059E-6	4.622E-8	4.37
mean value			3.93

measured. The precision of determination of the mean value was 0.13 % (absolute).

The precision can be improved with increasing the total number of particle in one sample. An important advantage of the parameterisation in determination of the fat content in milk is the absolute determination without the need for chemical calibration of the instrument. The conventional methods for fat determination are based on IR spectroscopy or turbidimetry and these methods require a homogenized sample.

4.5. RED BLOOD CELLS

The red cells, erythrocytes, are very numerous in the blood. Usually, they measure 6.6–7.5 μm in diameter. However, cells with a diameter higher than 9 μm (macrocytes) or lower than 6 μm (microcytes) have been observed. These cells are non-nucleated (among vertebrates, only the red cells of mammalians are lacking a nucleus), biconcave discs that are surrounded by a thin, elastic membrane and filled with hemoglobin. They are soft, flexible and elastic and therefore move easily through the narrow blood capillaries. The primary function of these cells is to carry oxygen from the lungs to the body cells.

The erythrocytes played an important role in verification of solutions of direct light scattering problem of non-spherical particles because of simple internal structure and stable biconcave discoid shape. At present the red blood cells form the size border where effective methods of simulation of light scattering can be applied. Tsinopoulos and Polyzos (1999) performed the study with a goal to solve the scattering problem of an undeformable, averagesized red blood cell (RBC) illuminated by a He–Ne laser beam and, for various orientations of the RBC with respect to the direction of the incident light, to evaluate the scattering patterns in the forward, sideways, and backward directions.

4.5.1. Optical model

A mature erythrocyte has a biconcave discoid shape (Figure 4.13). A red blood cell (RBC), erythrocyte, is composed of hemoglobin (32 %), water (65 %), and membrane components (3 %) and does not contain any nucleus (Mazeron at al, 1997). The shape of the erythrocyte is described by Skalak et al (1973)

$$z^2 = (0.86d/2)^2(1 - (2x/d)^2)(0.01384083 \\ + 0.2842917(2x/d)^2 + 0.01306932(2x/d)^4). \quad (4.2)$$

In equation (4.2) is described the erythrocyte profile in cross section with $= 0$, where x, y, z are Cartesian coordinates and d is the maximum axis length of the erythrocyte. The erythrocyte maximum axis length varies from 6 to 8 μm. The real part of the refractive index at $\lambda = 0.6328$ μm falls between 1.40 and 1.42 and is mainly caused by haemoglobin, 1.615, and water, 1.333 (Mazeron at al, 1997). The erythrocyte is surrounded by a thin lipid bilayer (Tycko et al, 1985) having a thickness of 7 nm, which is not incorporated into our model. In this study we also assume that the imaginary part of the refractive index is negligible at the above-mentioned wavelength λ.

Important parameters in the optical model of a particle are the size parameter, $\alpha = \pi d m_0/\lambda$, and the relative index of refraction, $m = m'/m_0$, where λ is the wavelength of the incident radiation, d is the maximum size of the erythrocyte, and m' and m_0 are the refractive indices of the particle and the medium, respectively. A size parameter and a relative index of refraction for an erythrocyte of 42 ($d = 6.3$ μm) and 1.058 ($m' = 1.41$), respectively, were used in our calculations.

In general, the intensity of light scattered by a particle depends not only on the size and the refractive index of the particle but also on its shape and orientation in relation to the direction of incident light. These factors complicate the interpretation of light-scattering data and reduce the accuracy and precision of particle analysis. Because a native red blood

Chapter 4. Applications

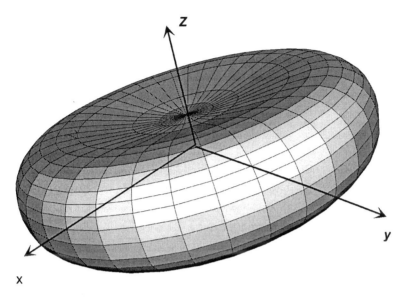

Figure 4.13. Model of a red blood cell. The shape of the red blood cell is described in equation (4.2)

cell has a biconcave discoid shape, the analytical method used in practical haematology, eliminates the shape and orientation dependence of the cells by isovolumetric sphering of the cells. The sphered cells are essentially homogeneous dielectric spheres. Therefore, with respect to light scattering, a sphered RBC can be characterized by size and complex index of refraction $n' = n'_R - in'_I$. Because the interior of the cell is almost completely occupied by water and hemoglobin, the variations in refractive index n' between cells can be attributed solely to variations in hemoglobin concentration. This is described by the following empirical equation:

$$n'_R - n_0 = \beta \times HbC, \tag{4.3}$$

where β is the coefficient expressed in dl/g and n_0 is the refractive index of the surrounding medium. The radiation wavelength interval of interest in this investigation is in the 0.5–1.2 μm range. Within this wavelength range the coefficient $.\beta$ has a typical value of 0.0019 dl/g and the hemoglobin is the only cell constituent exhibiting a significant amount of absorption. Consequently, assuming that Beer's law applies, the imaginary part of n is given by

$$n'_I = \frac{1}{4\pi} \lambda \sigma \times HbC \frac{N_A}{M}, \tag{4.4}$$

where σ is the absorption cross section of hemoglobin ($8.1 \cdot 10^{-18}$ cm^2 for

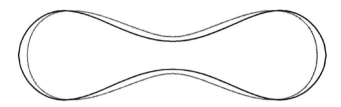

Figure 4.14. Red blood cell profiles calculated from equation (4.2) (gray) and equation (4.5) (black)

$\lambda = 0.6328 \cdot 10^{-4}$ cm), M is the molecular weight of hemoglobin (66.500), and N_A is Avogadro's number ($6.02 \cdot 10^{23}$). The typical hemoglobin concentration HbC of the erythrocyte (0.32 g/cm^3) gives the imaginary part of refractive index n'_I of $1.2 \cdot 10^{-4}$, which is in agreement with Hammer *et al* (1998). Thus, for a given wavelength, the complex index of refraction of the RBC is determined by equations (4.3) and (4.4) by a single physical variable, which is the cell hemoglobin concentration.

The equation (4.2) that describes the erythrocyte profile was improved with new experimental equipment (Fung et al, 1981).

$$z = 0.65\sqrt{1 - x^2}\left(0.1583 + 1.5262x^2 - 0.8579x^4\right). \qquad (4.5)$$

The equation (4.5) describes the red blood cell profile at $y = 0$. The profiles evaluated from equations (4.2) and (4.5) are shown in Figure 4.14 by gray and black lines, respectively. At the first stage of our study we applied the profile of equation (4.2) to simulate the light scattering of erythrocytes with WKB approximation (Section 1.4). Our latest results dealing with simulation of light scattering of erythrocytes from DDA relate to profile of equation (4.5).

4.5.2. Theoretical simulation

4.5.2.1. WKB approximation

The numerical calculations of light scattering for a spherical particle, having a diameter and a refractive index corresponding to those of the erythrocyte (diameter, $d = 6.3$ μm; refractive index, $m = 1.41$; wavelength, $\lambda = 0.6328$ μm; medium refractive index, $m_0 = 1.333$), and an exact solution derived from the Mie theory were compared. A modified version of the Mie code of Bohren and Huffman (1983) was used. This comparison was done to test the physical models that describe the light scattering of an erythrocyte.

The results of these numerical calculations of light-scattering LSPs for a sphere, which were carried out with Mie theory, RGD, WKB, and 2wWKB, are shown in Figure 4.15. We calculated the absolute scattering intensities

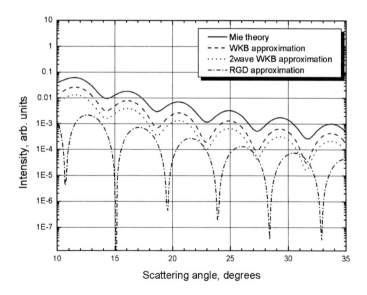

Figure 4.15. Light-scattering profiles of a sphered erythrocyte calculated from Mie theory, RGD, WKB, and 2wWKB approximations. Calculation parameters: diameter, $d = 6.3$ μm; refractive index, $m = 1.41$; wavelength, $\lambda = 0.6328$ μm; medium refractive index, $m_0 = 1.333$. The light-scattering profiles were normalized at 1 in the forward direction, polar scattering angle $\theta = 0$, and were shifted relative to one another

for all methods. Then all LSPs were normalized at 1 in the forward direction, with polar scattering angle $\theta = 0$. To observe easily the differences in the LSP shape, we shifted the LSPs relative to one another in Figure 4.15.

We can conclude that the maxima and the minima positions for the RGD approximation do not coincide with those of the WKB and Mie calculations, but the greatest deviation is in the visibility of the LSP. In the Section 3.1.2 we defined the LSP visibility as

$$V(15) = \frac{I_{\max} - I_{\min}}{I_{\max} + I_{\min}}, \qquad (4.6)$$

where $I_{\min}$ is the light-scattering intensity for the LSP minimum that occurs after the boundary angle at 15° and $I_{\max}$ is the light-scattering intensity for the next maximum. The LSPs calculated from the WKB and the 2wWKB approximations are coincident in the shape, and moreover the positions of the extrema are consistent with Mie theory. The main deviation in LSPs calculated from the Mie theory and the WKB approximation is observed in the visibilities. For example, the Mie and the WKB LSPs give visibilities of 0.416 and 0.585, respectively. This fact allows us to suppose that there is less visibility for experimental LSPs compared with the visibility of the LSPs calculated from the WKB approximation.

We calculated the erythrocyte LSPs, the scattering matrix element S_{11}, at different scattering angles varying from 10° to 35° by using the WKB approximation. The optical model of the erythrocyte described in Section 4.5.1 was used in calculations with a long axis length d of 6.3 μm and a relative refractive index of 1.058. Owing to the small thickness of the membrane, the membrane was excluded from the erythrocyte optical model. The calculated LSPs were normalized at 1 in the forward direction, $\theta = 0$, and were shifted relative to one another. The LSPs at different angular orientations β of the erythrocyte, where β is the angle between the axis of the erythrocyte symmetry and the direction of the incident beam, are shown in Figure 4.16. The orientation angle β was varied from 0° to 90° in 10° steps. The calculated LSPs show polymorphism.

4.5.2.2. Discrete dipole approximation

The progress in computing power and numerical methods has allowed us to apply the Discrete Dipole approximation for simulation of RBC light scattering. We used parallel super-computing algorithm described in Section 1.3. The optical model of the erythrocyte described in Section 4.5.1 was used in calculations with the profile described by equation (4.5). The erythrocyte

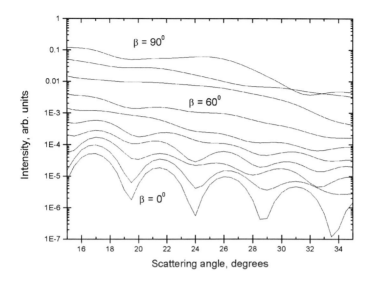

Figure 4.16. Light-scattering profiles of the erythrocyte calculated from the WKB for different orientations relative to the direction of the incident beam. The light-scattering profiles are normalized at 1 in the forward direction, polar scattering angle $\theta = 0$, and are shifted relative to one another

diameter and refractive index were varied from 6 μm to 9 μm and from 1.39 to 1.42, respectively. We assume that the imaginary part of the refractive index is negligible at the wave length λ of 0.6328 μm. The relative index of 1.333 for surrounded medium (saline) was used in our simulation.

The DDA gives a nice opportunity to make a choice of proper geometrical model of the erythrocyte to simulate light scattering. We have compared the LSP of particles shaped according to equation (4.5) and particles with the following shape geometry: oblate cylinder, oblate spheroid. The LSPs were compared for two orientations of the particles relating to direction of the incident beam, face-on and rim-on incidence. The parameters of the erythrocyte with a profile described by equation (4.5) are as follows: diameter $d = 7.60$ μm, maximal thickness of 2.89 μm, relative refractive index of 1.05. We have performed the comparison for the diameter-volume-equal particles. The LSPs of the erythrocyte and disk are shown in Figure 4.17. The shown LSPs allow us to conclude that erythrocyte can be modelled by a disk only for simulation of light scattering in the angular interval ranging from 0° to 15°.

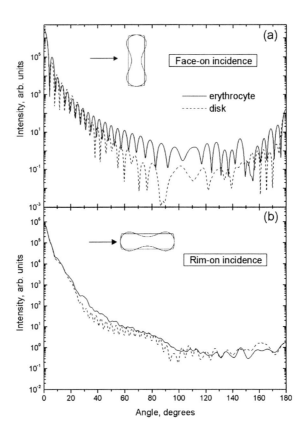

Figure 4.17. The light-scattering profiles of the erythrocyte and diameter-volume-equivalent disk; (a) face-on and (b) rim-on incidence

An oblate spheroid is the popular model in simulation of light scattering of individual erythrocyte. This model was used in simulation of light scattering with T-matrix method (Nilsson et al, 1998). We have calculated the LSPs of diameter-volume-equal spheroid in face-on and rim-on orientations of the oblate spheroid and erythrocyte. The result of calculation is shown in Figure 4.18. The shown LSPs allow us to conclude that erythrocyte can be modelled by an oblate spheroid over a wide angular interval only for rim-on incidence. This fact is in agreement with the boundary-element methodology applied to study of light scattering of red blood cell (Tsinopoulos and Polyzos, 1999).

Our simulation of light scattering of a mature erythrocyte has shown that the LSP is rather sensitive to erythrocyte shape and DDA must be used in

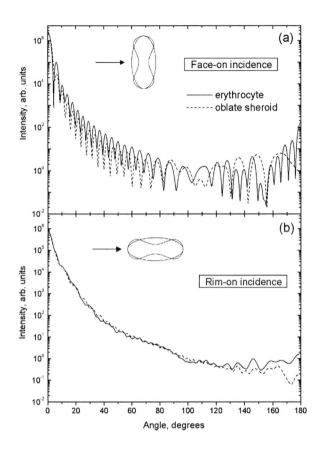

Figure 4.18. The light-scattering profiles of the erythrocyte and diameter-volume-equivalent oblate spheroid; (a) face-on and (b) rim-on incidence

a study of formation of the LPS under variation of erythrocyte characteristics. However the light scattering of the erythrocytes can be successfully simulated with the T-matrix method for the rim-on incidence. Fortunately the hydrodynamic system of the Scanning Flow Cytometer delivers mature erythrocytes into the testing zone in the proper orientation. This performance of the SFC gives a chance in solution of the ILS problem for mature erythrocytes with parameterisation or neural network because of T-matrix simulation requires less computing time substantially in comparison to DDA algorithm.

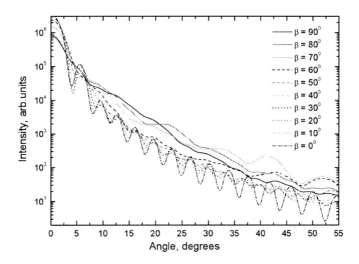

Figure 4.19. The light-scattering profiles calculated from discrete dipole approximation for 10 orientations of the single erythrocyte relating to the direction of incident light beam

Using DDA we studied an effect of orientation on LSPs of erythrocytes. The angular orientation β is the angle between the axis of the erythrocyte symmetry and the direction of the incident beam. We used the typical erythrocyte characteristics in calculation of light scattering that are as follows: diameter of 8.24 μm, aspect ratio of 0.328, volume of 112 μm^3, and relative refractive index of 1.05. The LSPs of ten orientations are shown in Figure 4.19. The orientation of the erythrocyte modifies the LSP substantially for instance middle scattering, angular region ranging from 15° to 60°, is reduced in intensity for symmetric orientation ($\beta = 0$) whereas the oscillating structure vanishes varying the orientation from $\beta = 0$ to $\beta = 90°$. The oscillating structure loss is in agreement with our WKB calculations shown in Figure 4.16. Consequently the middle light-scattering intensity and LPS structure can be a good indication on a mature erythrocyte orientation in flow cytometry.

From practical point of view we are interesting to study a sensitivity of LSP structure to variations of erythrocyte volume and diameter, the orientation corresponds to orthogonal directions of erythrocyte symmetry axis and incident beam. These requirements are caused by the orientating effect of the hydrofocussing head of the Scanning flow Cytometer where non-spherical

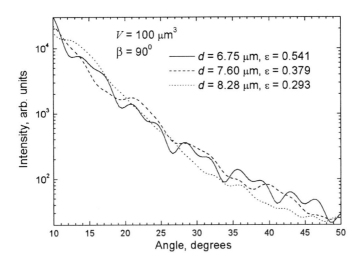

Figure 4.20. The light-scattering profiles of mature erythrocyte with different diameters

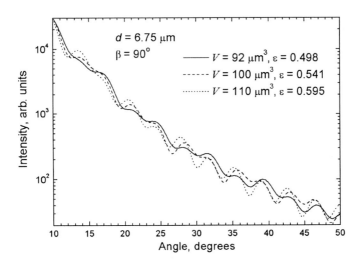

Figure 4.21. The light-scattering profiles of mature erythrocyte with different volumes

particles are oriented with a long axis along the flow line. We calculated the LSPs of erythrocyte with a typical volume of 100 fl varying the erythrocyte diameter. The results of DDA simulation are shown in Figure 4.20. The variation of the diameter does not change the intensity substantially whereas the visibility of oscillating structure is reduced with an increment of the erythrocyte diameter.

Additionally we varied the thickness of the erythrocyte that results to the variation in a volume with a constant diameter. The erythrocyte volume is the important haematological index measured with modern automatic haematology analyzers. Flow cytometry and Coulter's cell are the instrumental solutions utilized for measurement of erythrocyte volume distribution. The erythrocyte LSPs calculated for three volumes with a typical diameter of 6.75 μm are shown in Figure 4.21. The light-scattering intensity does no depend on variation of the erythrocyte volume or thickness. There is the same tendency in the LSP structure that is as follows: the visibility of the oscillating structure is increased when a particle loses a nonsphericity.

Our simulation of light scattering of mature erythrocytes has allowed us to conclude that the orthogonal orientation of an erythrocyte relating to direction of incident beam does not provide a sufficient sensitivity of the LSP to the erythrocyte characteristics. The coincidence of erythrocyte symmetry axis and direction of incident beam gives an advantage in solution of ILS problem because the oscillating structure of the LSP is very sensitive to variation of erythrocyte characteristics.

4.5.3. Experimental results

4.5.3.1. Mature erythrocytes

A sample containing approximately 10^6 cells/ml was prepared from fresh blood by use of buffered saline for dilution. We continuously measured 443 light-scattering profiles of erythrocytes, and 10 of them are shown in Figure 4.22. Then we chose each LSP shown in Figure 4.22 by sorting out those most similar to the corresponding LSP calculated from the WKB approximation shown in Figure 4.16. Polymorphism of the LSPs of individual erythrocytes was also observed in the experiment.

We are not interested in absolute light-scattering intensity in this study. To observe easily the differences in the LSP shape, we shifted the LSPs relative to one another in both Figure 4.16 and Figure 4.22. A qualitative agreement in the measured and the calculated LSPs is observed, but other differences raise questions. First, the experimental LSPs are more

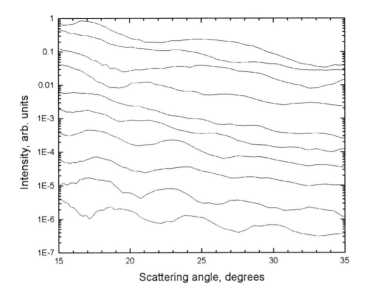

Figure 4.22. Experimental light-scattering profiles of the erythrocyte measured with the SFC. We shifted the light-scattering profiles relative to one another to observe the differences in the shape easily

polymorphic because of variations in erythrocyte length and thickness. We can conclude from the results in Figure 4.15 that an exact theory (the Mie theory) gives a lower visibility than the visibility calculated from the WKB approximation for the same particle. The largest visibility found in a set of experimental LSPs was always lower than the visibility of the calculated LSP derived from the WKB approximation.

The complete quantitative analysis of measured LSPs of erythrocytes is not available because of the complexity of the exact light-scattering theory for nonspherical particles, resulting in the long computational times needed to extract the erythrocyte parameters from the light-scattering data. The WKB approximation also requires long times for calculation of the $S11$ element of the scattering matrix. Below we have a preliminary analysis of the measured LSPs to clarify the orientation properties of the hydrofocusing system of the SFC. Only those LSPs that correspond to an orientation angle β of 0, 10, 20, and 30° have a clear and characteristic oscillating structure according to the WKB approximation (Figure 4.16). The visibility $V(15)$

of the mentioned LSPs obtained from WKB computations varies from 0.7 to 0.071. We found that the visibility $V(15)$ of 208 LSPs in our experiment was higher than 0.07 and corresponds to 47 % of the total amount of the measured erythrocytes. This value can be considered as the lowest estimation of the visibility because the WKB approximation provides enlarged visibilities compared with the exact theory. These computations indicate that only 7 % of the erythrocytes are oriented at angles between 0 and 30°. The reason for this disagreement is probably due to hydrodynamic effects on the nozzle or a funnel effect in the inlet tubes of the hydrofocusing cavity of the SFC. The motion of the erythrocyte within the SFC seems to be nonstationary, and the preorientation of the erythrocyte within the hydrofocusing cavity plays an important role in the use of SFC for analysis of the light-scattering properties of nonspherical particles.

An availability of DDA codes realized with parallel super-computing processors has allowed us to understand the hydrodynamic effects on the nozzle. As the WKB approximation give sufficient errors in evaluation of the LSP visibility we have to apply the DDA codes to compare the experimental LSP visibility distribution and LSP visibility calculated from DDA for different orientations of the erythrocyte. Practically the DDA can not be used to fit the experimental and theoretical LSPs because the DDA calculation requires approximately 10 hours for fixed erythrocyte orientation even with the super-computer.

A sample containing approximately 10^6 cells per ml was prepared from fresh blood by use of buffered saline for dilution. We continuously measured 3000 LSPs of erythrocytes with the SFC. We compared the LSPs of mature erythrocytes calculated from DDA for typical characteristics of single erythrocyte in different orientations (Figure 4.19). The comparison performed with χ^2-test has allowed us to make the following statement: there is no erythrocyte with orientation angle less than 60 degrees in testing zone of the SFC.

In order to determine the distributions of erythrocyte characteristics in the blood sample measured experimentally we calculated the LSPs of erythrocytes with the characteristics shown in Table 4.3. The LSPs of the erythrocyte with a fixed aspect ratio were calculated for orientation angle ranging from 60° to 90° with a step of 10°. A set of 92 LSPs was computed with the DDA method.

Each LSP from the set was compared with 3000 LSPs with the χ^2-test. The erythrocyte characteristics of the theoretical LSP were bound to the erythrocyte measured experimentally if the χ^2-value was less than

Table 4.3. The parameters of erythrocytes calculated for statistical analysis. There are values of aspect ratio ε in cells of table

volume, mm^3 \ diameter, mm	6.08	6.33	6.51	6.75	6.84	7.01	7.60	8.28
86	0.636	0.564	–	0.464	–	–	0.328	–
92	–	0.604	0.554	0.498	–	–	0.348	–
100	–	–	0.603	0.541	0.519	0.482	0.379	0.293
105	–	–	–	0.568	–	–	0.398	0.307
110	–	–	0.663	0.595	0.577	0.444	0.417	0.322

the threshold. There are a few results of the χ^2-test in Figure 4.23. The threshold was determined from the comparison of the mean cell volumes retrieved from the χ^2-test and ordinary haematological analysis of sphered erythrocytes (see next section).

The χ^2-test has allowed us to build up distributions of mature erythrocyte characteristics measured with the Scanning flow Cytometer. The distributions were formed with 10 % of LPSs measured. We lost 90 % of erythrocytes because the χ^2-value that exceeds the threshold. The loss was caused by incompleteness of the LSP set calculated from DDA method including a variation of haemoglobin concentration of erythrocytes that results to the variation of the erythrocyte refractive index. The retrieved distributions are shown in Figure 4.24.

The measured distribution of orientation angle of erythrocytes within the testing zone of the SFC is in agreement with previously made statement that assumes a negligible amount of erythrocytes with orientation angel less than 60°. The mean mature erythrocyte diameter is in good agreement with literature (Fung et al, 1981).

4.5.3.2. Sphered erythrocytes

The LSPs of sphered erythrocytes can be analyzed with the parametric solution described in Section 3.2. For example, this method applied to the erythrocyte LSP in Figure 4.25b gave a size and a hemoglobin concentration in the erythrocyte of 5.92 μm and 60.1 g/dl, respectively. The same LSPs were also processed with a nonlinear fitting method where a measured LSP was fitted by a LSP calculated from Mie theory. The result of the

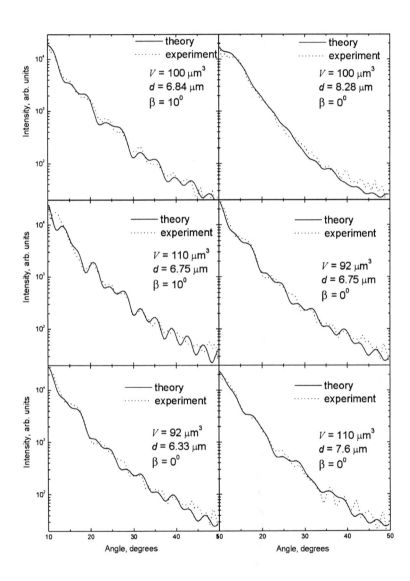

Figure 4.23. Experimental and theoretical light-scattering profiles of individual mature erythrocytes

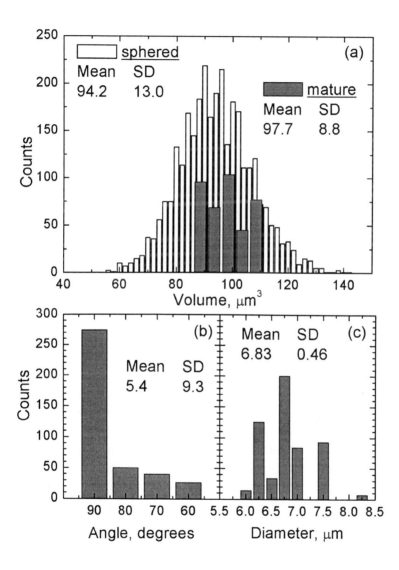

Figure 4.24. (a) The erythrocyte volume distributions of the blood sample retrieved from analysis of mature and sphered erythrocytes. (b) The orientation angle and (c) diameter distributions of mature erythrocyte

fitting is shown in Figure 4.25b (solid curve). There was no disagreement between the methods in determining the erythrocyte size. The difference in erythrocyte hemoglobin concentration is caused by the difference in visibility (Figure 4.25b). The visibility of the theoretical LSP exceeds the visibility of the experimental LSP, which results in a smaller hemoglobin concentration (Figure 3.16). The parameterisation with equations (3.19) and (3.20) permits determination of the erythrocyte characteristics in real time, whereas the best-fit method requires approximately 5 min to retrieve the erythrocyte size and the hemoglobin concentration with a Pentium II processor. The substantial advantage for both methods is the independence from absolute scattering intensities. Consequently, these methods do not require calibration of the SFC with certified particles.

RBC's in a patient sample are characterized by measuring their cell volume distribution and distribution of cell hemoglobin concentration. An isovolumetric spherization has to be used to measure these distributions of sphered erythrocytes. Water does not provide the isovolumetric spherization of erythrocytes. Hence we can characterize the RBC's by the volume distribution of sphered erythrocytes and hemoglobin weight distribution. To retrieve these distributions, we measured the sphered erythrocyte sample with the SFC. The measured LSPs were processed by the parameterisation described in Section 3.2. The results of the analysis are presented as erythrocyte hemoglobin weight versus an erythrocyte volume map (Figure 4.26a) and histograms (Figure 4.26b, c).

At present most blood cell analyzers are based on ordinary flow cytometry and the two-angular light-scattering method to determine the red cell parameters by using isovolumetric spherization of erythrocytes as described by Kim and Ornstein (1983). Obviously isovolumetric sphering can also be applied to analyzing RBC's with scanning flow cytometry and the parameterisation. We believe that isovolumetric sphering will give more stable and reproducible results with this approach. The RBC analysis will be simplified by the removal of calibration of the instrument with certified particles.

In this study we have developed a method for simultaneous measurement of RBC volume and hemoglobin concentration where calibration of the instrument is not necessary. Instrumentally and theoretically this method is based on a SFC and on a parametric solution of the inverse light-scattering problem, respectively. We have applied an algorithm that allows construction of approximating equations and that was previously developed for analysis of homogeneous spherical particles, to construct the equations relating the LSP parameters to the absorbing particle characteristics. The method

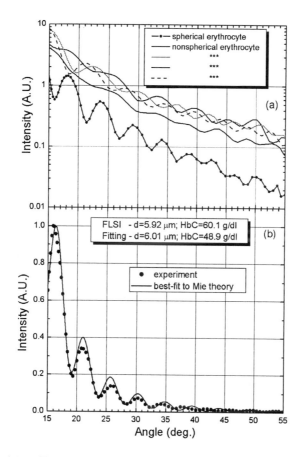

Figure 4.25. Experimental light-scattering profiles of (a) individual nonspherical and spherical erythrocytes and (b) a spherical erythrocyte processed with a parameterisation and with a fit of Mie theory

has been tested by the measurement of sphered erythrocyte volume and hemoglobin concentration. However, the algorithm can be easily used to construct the approximating equations for other types of absorbing particles, although one has to take into account the coefficient β of equation (4.3) and the absorption cross section σ of equation (4.4).

The particle size and concentration of absorbing substance can be determined in real time with the developed method. At present, the SFC operates at a rate of 500 particles/s, but the rate can be increased by using specially developed electronics and software as well as a more powerful processor.

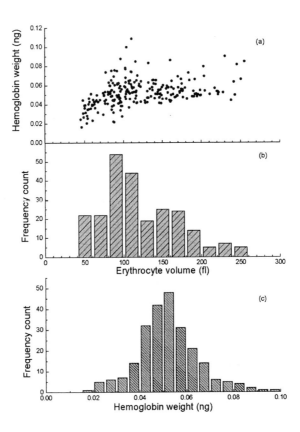

Figure 4.26. Analysis of a blood sample containing sphered erythro-cytes. The results of the analysis are shown as (a) the erythrocyte hemoglobin content versus the volume map, histograms of the distribution of (b) the erythrocyte volume, and (c) the hemoglobin content

4.6. BACTERIAL CELLS

Bacteria are a major component of the unseen world and play a decisive role in the maintenance of life on this planet. *Escherichia coli* (the colibacillus, *E.coli*), which normally constitutes 80 % of the aerobic flora in the digestive tube, has the rod-like shape characteristic of bacilli. The cell is a common bacterium in microbiology.

4.6.1. Theoretical simulation

We have made a study of light-scattering properties of individual *E.coli* cells. Calculations of electromagnetic scattering from particles of arbitrary shape may be exactly performed with the T-matrix method. We modelled the *E. coli* cell by a prolate spheroid. The laser wavelength of 488 nm and medium refractive index of 1.333 were used in the calculations. The intensity of light scattered by a particle can be described via matrix formalism with the following expression:

$$I_s(\theta,\varphi) = I_i \, (1,0,0,0) \begin{pmatrix} S_{11} & S_{12} & S_{13} & S_{14} \\ S_{21} & S_{22} & S_{23} & S_{24} \\ S_{31} & S_{32} & S_{33} & S_{34} \\ S_{41} & S_{42} & S_{43} & S_{44} \end{pmatrix} \begin{pmatrix} 1 \\ \cos(2\varphi) \\ \sin(2\varphi) \\ 0 \end{pmatrix}. \quad (4.7)$$

The Stokes four vector on the right describes the linear polarized light with angle φ relative to the lab system; I_t is the intensity of incident beam; $S_{ij} = S_{ij}(\theta,\varphi)$ are the elements of the scattering matrix; $(1,0,0,0)$ is the four vector that describes the photo detector. Evaluating the matrix multiplication and integrating over azimuthal angle range of $0°$–$360°$, the intensity of light measured by the SFC is as follows:

$$I_s(\theta) = \int_0^{2\pi} I_s(\theta,\varphi)\,\mathrm{d}\varphi = I_i \int_0^{2\pi} (S_{11} + S_{12}\cos(2\varphi) + S_{13}\sin(2\varphi))\,\mathrm{d}\varphi. \quad (4.8)$$

We calculated the S_{11} element for a prolate spheroid with the following parameters: a diameter d of 0.7 μm; rotational axis length l of 2.1 μm; a relative refractive index of 1.048; an angle of 10° between the direction of incident light and rotational axis of the spheroid. The diameter, length, and relative refractive index are in the range of variation of these parameters of *E.coli* cells (Bronk et al, 1992). The S_{ij} element calculated in polar angle range of 10°–60° is shown in Figure 4.27 in spherical coordinates $(\log(I_s), \theta, \varphi)$. The incident beam is directed from the center of lower face of the axis box to the center of the upper face. Earlier we defined the LSP visibility as follows: $V(\theta_b) = (I_{\max} - I_{\min})/(I_{\max} + I_{\min})$, where $I_{\min}$ is the light-scattering intensity for the minimum that occurs after the boundary angle θ_b and $I_{\max}$ is the light-scattering intensity for the next maximum. Let us set the boundary angle of 15°. The three dimensional LSP presented in Figure 4.27 allows us to conclude that an integration of LSP over azimuthal angles decreases the LSP visibility substantially. We measured in our experiments the LSP visibility of *E.coli* cells that was much larger in comparison to the LSP visibility calculated from the T-matrix method for the cell with 10° oriented relative to incident beam and the visibility was comparable with 0°-oriented cells. This fact is in an agreement with the statement that the hydrodynamic system of the SFC orients spheroidal particles (Kachel et al, 1990). The flow line coincides with the long axis of the spheroid.

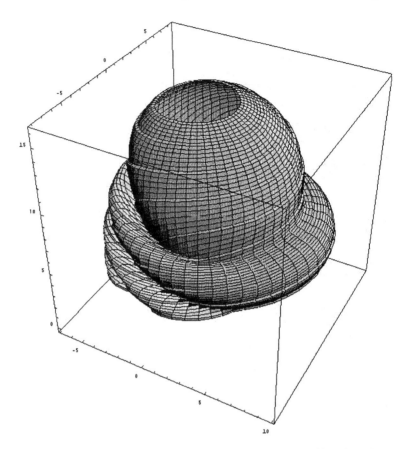

Figure 4.27. Three dimensional-light-scattering profile of prolate spheroid shown in spherical coordinates $(\log(I_s), \theta, \varphi)$ where I_s is the scattering intensity, and θ and φ are the polar and azimuthal angles of the spherical coordinates, respectively

To simulate the scattering, the cells were modelled by prolate spheroids. Bronk et al. (1995) modelled these cells as a cylinder capped with hemispheres of the same radius as the cylinder and applied the coupled dipole approximation to calculate the scattering from such particles. This model is based on a microscopic analysis of *E.coli* cells and may be more appropriate for the scattering simulation comparing the prolate spheroid. The light-scattering calculations helped us to find that the *E.coli* cells are oriented by Poiseuille profile flow and position of the LSP minimum moves to the forward direction with increasing of a length of the rotational axis of the prolate spheroid (Figure 4.28).

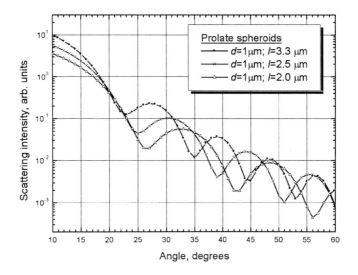

Figure 4.28. The light-scattering profiles of prolate spheroids calculated from the T-matrix method. The calculation was performed with a relative refractive index of 1.048. The direction of incident light is coincided with the rotational axis of the spheroid

4.6.2. Experimental verification

The SFC allows the measurement of absolute light-scattering characteristics of particles. It was shown in Section 4.1 that the light-scattering LSP of a polystyrene particle measured with the SFC gave a good agreement with the LSP calculated from the Mie theory. The size and refractive index of spherical homogeneous particles were determined fitting these LSPs to each other. The differential cross section characterizes the efficiency of particle to scatter light (Bohren and Huffman, 1983). To determine the differential cross section of *E.coli* cells, we mixed the polystyrene particles and cells. The mixture was measured with the SFC and results of the measurements are presented in Figure 4.29. The pointers show the sets of the LSPs of monodisperse polystyrene particles (0.9 μm and 3 μm) and *E.coli* cells. The differential cross section scale was determined from a fit of the experimental LSP of polystyrene particles and theoretical LSP calculated from the Mie theory.

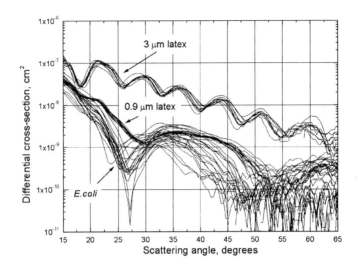

Figure 4.29. The differential cross-section of *E.coli* cells and polystyrene particles measured with scanning flow cytometer

We have also analyzed the LSPs of *E.coli* cells in different phases of population evolution. In particular, the cells in logarithmic and stationary phases prepared as described above, have been measured with the SFC. The set of these LSPs is shown in Figure 4.30a, b, respectively. The main difference has been found in the position of the first minimum of the LSPs. The corresponding distribution of minimum position for both phases is shown in Figure 4.31.

The above-presented results demonstrate a performance of the SFC in an analysis of light-scattering properties of individual *E.coli* cells.

We have determined the absolute scattering cross section of individual *E.coli* cells using an ability of the SFC to precisely measure parameters of homogeneous spherical particles. The amount of light scattered by an *E.coli* cell in the polar angular range from 15° to 65° can be easily estimated from the LSPs presented in Figure 4.29. These data allow one to design an ordinary flow cytometer properly to classify measured particles into different categories. For instance, this mixture of particles will be effectively sorted with the measurement of forward scattering in the angle range of 23°–26° and will give an ambiguous result if the forward light scattering will be measured in the angle range of 17°–19°. Moreover, the required sensitivity of photodetector can be estimated from the differential cross section.

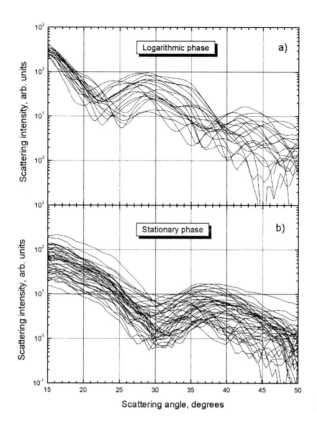

Figure 4.30. The experimental light-scattering profiles of *E.coli* cells in logarithmic (a) and stationary (b) phases

The LSP of the *E.coli* cell is rather sensitive to the phase of cell evolution. Two main differences have been found for cells in logarithmic and stationary phases of growth (Figure 4.30). First, the LSP minimum is located around 22° for logarithmic phase and around 30° for stationary phase (Figure 4.31). Hence, the location of the LSP minimum can be used to identify the phase of growth of *E.coli* cells. For example, the cells measured in the mixture with polystyrene particles were in a logarithmic phase (Figure 4.29). Moreover, we are able to state that these cells are older than the cells presented in Figure 4.30a because of minimum position around 26°. According to our calculations shown in Figure 4.28, we may assume that a length of the cells is reduced in evolution. This fact is in agreement with microscopic analysis made by Bronk et al. (1992). Second, the variations of scattering

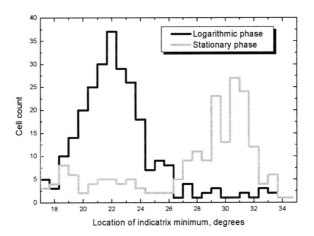

Figure 4.31. The distribution of the location of the light-scattering profile minimum for *E.coli* cells in logarithmic (black line) and stationary (gray line) phases of growth

intensity in angles around 15° are smaller for cells in the logarithmic phase compared to the stationary phase (Figure 4.30a, b). The large dispersion in scattering intensity for the cells in stationary phase may be caused by the large difference in the internal structure of the cells. This assumption is in an agreement with an analysis of *E.coli* cells in different growth phases with electron microphotographs (Krossbacher et al, 1998).

The performed study of light-scattering properties of individual *E.coli* cells has demonstrated the performance of SFC in measuring the angular light-scattering function of cells. The measured differential cross section of *E.coli* can be the first contribution into our database of the light-scattering function of different microorganisms. One LSP occupies a storage space of approximately 2 kb and LSP database can be easy stored on an ordinary hard disk. With additional studies of light-scattering functions of other cells and with an enlargement of the database, microorganisms presented in the database can be identified by applying modern powerful mathematic methods and computers. Moreover, the database should consist of light-scattering functions of cells in a logarithmic phase of growth that shows more stable LSPs of individual particles.

The other important problem is determination of spheroidal modeled particle characteristics (length, diameter, and refractive index) from the

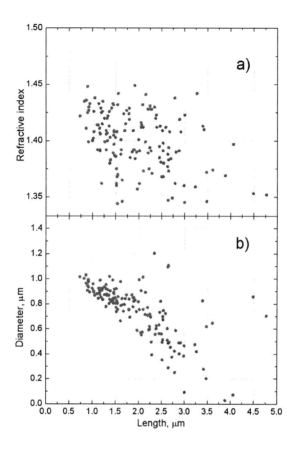

Figure 4.32. The refractive index versus length (a) and diameter versus length (b) maps retrieved from experimental light-scattering profiles of *E.coli* cells with the parametric solution of the inverse light-scattering problem for prolate spheroids

LSP measured. These characteristics can be determined from the parametric solution of the inverse light-scattering problem.

The SFC allows the measurement of the distribution of cell parameters instead of measurements of mean parameters of cell populations (as is available with other light-scattering techniques). This technique seems like an ideal method for investigating individual cell-parameter changes during a cell cycle. A careful study of evolution of distributions of cell parameters is a goal of our future research.

The equations (3.24)–(3.26) can be applied to determine the prolate spheroid characteristics from experimental LSPs. We have processed the experimental *E. coli* LSPs and the results are shown in Figure 4.32. The cell characteristics determined from the parametric solution of the ILS problem are in agreement with literature. The algorithm introduced with this work gives an ability to control the cell characteristics in suspension dispersed in a selective broth.

4.7. LYMPHOCYTES

White blood cells which include granular and agranular cells are an important part of the body's immune system, helping to destroy invading microorganisms. The lymphocyte is an agranular cell with very clear cytoplasm. Lymphocytes are quite common in the blood: 20–40 %, 8–10 μm in diameter and generally they are smaller than the other white blood cells but they are still larger than erythrocytes. The cytoplasm is transparent. Lymphocytes are distinguished by having a nucleus which may be eccentric in location, and a relatively small amount of cytoplasm. The small ring of cytoplasm contains numerous ribosomes. The nucleus is round and large in comparison to the cell and it occupies most of it. In any case, some of the cytoplasm remains visible, generally in a lateral position. According to the quantity of cytoplasm, lymphocytes are divided into small, medium and large. These cells play an important role in human immune response. The T-lymphocytes act against virus infected cells and tumour cells. The B-lymphocytes produce antibodies.

Ordinary flow cytometry was successfully applied to discriminate the white blood cells. Granulocytes, monocytes, lymphocytes occupied unique positions in the two-dimensional space created by the forward and side light-scattering intensities. de Grooth et al (1987) proposed measurement of depolarization of side light scattering for discrimination inside of granulocyte subpopulation. Measurement of the depolarized orthogonal light scattering in flow cytometry enables one to discriminate human eosinophilic granulocytes from neutrophilic granulocytes (Terstappen et al, 1987). Finally they measured forward light scattering, orthogonal light scattering, and the fluorescence intensities of unlysed peripheral blood cells, labeled with CD45-phycoerythrin and the nucleic acid dyes LDS-751 and thiazole orange utilizing an ordinary flow cytometer. Erythrocytes, reticulocytes, platelets, neutrophils, eosinophils, basophils, monocytes, lymphocytes, nucleated erythrocytes, and immature nucleated cells occupied unique positions in the five-dimensional space created by the listmode storage of the five independent parameters (Terstappen et al, 1991).

In this study we analyzed the light scattering of most important subpopulation of white blood cells — lymphocytes. Formation of the lymphocyte LSPs was studied with the aim of development of appropriate optical model of a lymphocyte to solve the ILS problem for their characterization. The light scattering of lymphocytes was simulated by means of an algorithm that allows calculations of scattering matrix of two concentric spheres with the following characteristics: d and n_c are the diameter and refractive index of the inner sphere; D and n_s are the diameter and refractive index of the outer sphere, respectively. Additionally we were able to simulate lymphocyte light scattering with the multi-layered model defined by the layer diameter d_i and layer refractive index n_i where i is the number of layer. The choice of the models is based on analysis of experimentally measured LSPs of individual lymphocytes.

A sample that contains $5 \cdot 10^5$ lymphocytes was analyzed with the Scanning flow Cytometer. The measured LPSs were modified by means of equation (3.12) and spectral decomposition was applied to the modified LPSs. Contrary to a homogeneous particle with one-peak spectrum, we observed a few peaks in the spectrum of single lymphocyte. This fact is in agreement with results introduced in Section 3.3. The modified LSPs of two single lymphocytes with their spectral decomposition are shown in Figure 4.33 by points. To retrieve the lymphocyte characteristics form the measured LPS we applied the solution of the ILS problem for two concentric spheres introduced in Section 3.3. The results are as follows: (a) lymphocyte characterized by $d = 6.66$ μm, $n_c = 1.440$, $D = 9.17$ μm, $n_s = 1.358$; (c) lymphocyte characterized by $d = 6.38$ μm, $n_c = 1.456$, $D = 8.74$ μm, $n_s = 1.355$. The lymphocyte characteristics retrieved were used to calculate the modified LSP and spectral decomposition. These modified LSPs with spectral decomposition are shown in Figure 4.33 by grey lines. There is substantial disagreement between theory and experiment in the figure although the location of the maximal peak in the spectrum correlates with the diameter of inner sphere. This disagreement is probably caused by the insufficient two-layer model of single lymphocyte.

In order to clarify the reason of disagreement of the theoretical and experimental results we modelled a lymphocyte by a five-layered sphere. The experimental LSPs shown in Figure 4.33 were fitted by particles formed by five layers. The resulting particles can be characterized by the following parameters: (a) lymphocyte characterized by $d_1 = 1.468$ μm, $n_1 = 1.402$,

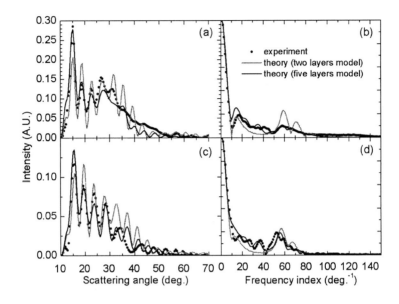

Figure 4.33. (a), (c) The modified light-scattering profiles and (b), (d) spectral decomposition of two single lymphocytes measured with scanning flow cytometer. The experimental data is marked by points whereas the theoretical fits are marked by grey and black lines for two and five layers models, respectively

$d_2 = 6.264$ μm, $n_2 = 1.478$, $d_3 = 7.895$ μm, $n_3 = 1.368$, $d_4 = 8.650$ μm, $n_4 = 1.360$, $d_5 = 9.518$ μm, $n_5 = 1.342$; (c) lymphocyte characterized by $d_1 = 1.418$ μm, $n_1 = 1.404$, $d_2 = 5.975$ μm, $n_2 = 1.465$, $d_3 = 6.606$ μm, $n_3 = 1.425$, $d_4 = 8.918$ μm, $n_1 = 1.366$, $d_5 = 9.708$ μm, $n_5 = 1.342$. The corresponding LSPs with spectral decomposition are shown in Figure 4.33 by black lines. The five-layered model has evidently given the better agreement between experiment and theory. On the other hand we have found that the central part of lymphocyte has smaller refractive index comparing the next layer.

The theoretical study of scattering of multi-layered spheres and experimental measurement of LSPs of individual lymphocytes has allowed us to conclude that a lymphocyte could not be modelled by two concentric spheres. The multi-layered sphere or eccentric spheres models must be apply to simulate light scattering of individual lymphocytes.

4.8. BISPHERES

Agglutination is one of the oldest serological reactions that results in clumping of a cell suspension by a specific antibody directed against an antigen (van Oss, 2000). Nowadays this phenomenon is widely used in different applications. Although there are a number of high sensitive and resourceful techniques and methods that have been developed for diagnostics purposes, the interest to rapid, simple and inexpensive tests still remains (Peula-Garsia et al, 2002). Antigen-coated latex is used for quantifying antibody concentration in a specimen (passive agglutination), and antibody-coated latex for antigen concentration (reversed passive agglutination). The enhanced latex particle immunoagglutination test is widely used in the diagnostics of various infections and to detect biomarkers or some chemical compounds in biological fluids (Ellis et al, 2000). On the other hand, agglutination takes place in different biological processes, such as platelets or red blood cell aggregation (Long et al, 1999). The interest to agglutination is also supported by the general study of coagulating systems. We have studied a performance of the Scanning Flow Cytometer in identification of agglutinating particles.

In order to verify the performance in discrimination of clusters of spheres, the preliminary experiments with uncovered fluorescent latex particles were carried out. The particle sample was not sonicated. A non-specific aggregation of particles, resulted in formation of clusters of two and more particles, occurred in these conditions. Thus, single spheres ('monomers'), clusters of two spheres ('dimers'), and so forth were observed in measured sample. Both LSPs and fluorescence signals were recorded simultaneously for each measured particle.

The experimental signals measured by the Scanning Flow Cytometer from different particles are presented in Figure 4.34. The SFC signal consists of three separate parts. The first part is the light-scattering trace from a latex particle. The second part is the trigger pulse, which occurs when the particle passes a certain place (trigger position) in the testing zone. The third part is the fluorescence signal. The intensity of the fluorescence pulse is proportional to the number of dye molecules in measured particle. Thus, the intensity of fluorescence signal can be used to determine dimers, trimers and other clusters of fluorescent particles. The LSPs evaluated from the corresponding SFC traces of the corresponding particles are presented in Figure 4.35. One can see from Figure 4.34 and Figure 4.35, that LSPs can be used for monomer, dimer and higher multimers identification.

The size and refractive index of homogeneous spherical particles are evaluated from the LSP using the parametric solution of the ILS problem.

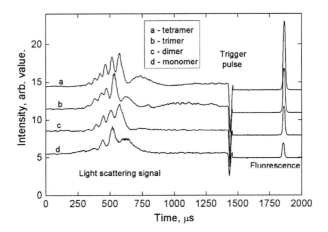

Figure 4.34. Native SFC traces of single particles from the sample with uncovered non-sonicated fluorescent latex particles. The signals are shifted on intensity for convenient view

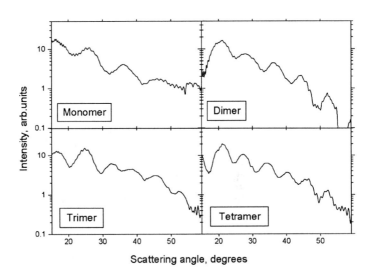

Figure 4.35. Light-scattering profiles for the particles as depicted in Figure 4.34. Native SFC traces from Figure 4.34 were transformed into angular dependency using the transfer function

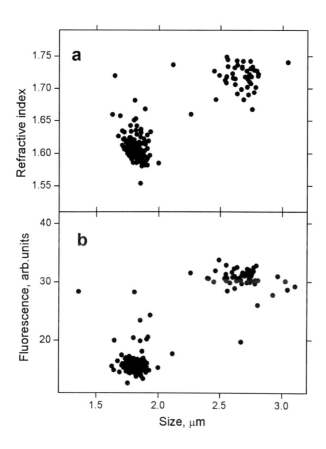

Figure 4.36. Parameters of particles from the sample containing non-sonicated fluorescent latex spheres. Each point represents a single measured particle: (a) refractive index vs. size map, the parameters were obtained from light-scattering profiles with the parameterisation; (b) fluorescence vs. size map

The particle characteristics obtained from measured LSPs are shown in Figure 4.36a (refractive index vs. size) and Figure 4.36b (fluorescence vs. size). Each point corresponds to one particle with its fluorescence, size and refractive index in Figure 4.36. It should be noted that the parameterization used in this work can be applied only for spherical particles, and it gives an effective size and refractive index for dimers. The particles were attributed to two groups: the left group consists of single spheres confirmed by their size (1.8 μm that is in agreement with manufacturer specification), and the right group corresponds to the clusters of two spheres. Good correlation between the obtained size and the fluorescence intensity leads to the conclusion that SFC is able to discriminate dimers from monomers from light-scattering profile.

These experiments have shown that "monomers" and "dimers" can be discriminated by means of parametric solution of the ILS problem that applicable for spherical particles.

Conclusion

In this work we have summarized the results relating to development of new experimental and theoretical approaches in characterization of individual particles from light scattering. Characterization assumes experimental determination of physical particle characteristics that requires a solution of the inverse light-scattering problem. In order to solve the inverse problem researchers have to develop an instrument which allows measurement of sufficient amount of light scattering data and to apply available solutions of the direct light-scattering problem for simulation. An experimental verification of the developed solution of the inverse problem with artificial and natural samples containing particles has to be performed. These matters have formed the structure of this scientific presentation.

The most important solutions of the direct light-scattering problem such as Mie theory, T-matrix method, Discrete Dipole approximation, Wentzel–Kramer–Brillouin approximation have been utilized in our research. We applied these theories to simulate light scattering of spherical and non-spherical particles, particles with inclusions. Mature and spherized erythrocytes, lymphocytes, milk fat particles, *E.coli* cells, and immunoactive polymer beads have formed a pool of bioparticles studied from the solutions of the direct light-scattering problem. The results introduced here have posed a few problems for next generations of theoretical consideration of light-scattering phenomenon. Although the Discrete Dipole Approximation provides simulation of light scattering of an arbitrary particle shape and structure the calculating time of available programming codes remains problematic for the particle volume-equal size large than $5\ .\mu m$. For instance, our simulation of light scattering of single mature erythrocyte, the volume-equal size of $6\ .\mu m$, for fixed orientation took time of 6 hours. Development of effective algorithms for light-scattering simulation of these particles is a vital problem.

From the experimental point of view we have focussed our activity on development of new generation of flow cytometers with enhanced performance in measurement of light scattering of individual particles. Historically, flow cytometry gained ground in the area of biology and medicine. This technique allows an effective analysis of cells. However an ordinary flow cytometer measures insufficient light-scattering information to solve the inverse light-scattering problem even for spherical particles. The Scanning Flow Cytometry opens a way to light-scattering spectroscopy where the light-scattering profile represents a light-scattering spectrum of the individual particle. This new technique increases the amount of light-scattering information measured from a single particle substantially that allows one to develop new effective methods to solve the inverse light-scattering problem for individual particles. Additionally similar to the classical spectroscopy the databases of light-scattering spectra of different kinds of particles can be developed and used in particle identification. The performance of the next generation of light-scattering instrument allowing measurement of a few light-scattering profiles of single particles, Polarizing Scanning Flow Cytometry, gives a chance to characterize non-spherical particles in arbitrary orientation. A usage of the polarizing light-scattering information in development of a solution of the inverse light-scattering problem is a straightforward way for advanced light-scattering technology.

A solution of the inverse light-scattering problem is an important stage of the characterization of particles from light scattering. From practical point of view a solution should be insensitive to experimental noise, independent on absolute light-scattering intensities (calibration-free approach), performed in real-time. The nonlinear regression of experimentally measured and theoretically calculated data satisfies the requirements excepting the real-time function. However the nonlinear regression can be realized only for homogeneous spherical particles because of an increase of particle characteristics to be determined from. The alternative of the nonlinear regression is the parametric solution of the inverse light-scattering problem that satisfies the abovementioned requirements. The parametric solution based on analytical approximating equations which relate the particle characteristics to light-scattering parameters. The analytical equations enable real-time characterization of particles, which is important for on-line work with a cell sorter. Additionally the use of particle defined light-scattering parameters makes absolute sizing possible without the need for calibrating electronic and optics tracks of a flow cytometer with standard particles. Moreover, the particle parameters determined with the parametric solution can be used as initial fitting parameters for non-linear regression to a solution of the direct light-scattering problem. At present the parametric solution of the inverse light-scattering problem allows one to determine optical characteristics of homogeneous spherical particles, spherical particles with absorbing inner substance, oriented spheroidal particles. There is no alternative of the parametric solution for determination of parameters of oriented prolate spheroid. The neural network approach demonstrated the promising results in solution of the inverse problem in scattering and we are going to expand an applicability of this method on particles of different shapes.

Development of the assays which use the above mentioned advantages of the Scanning Flow Cytometry is a vital problem. The SFC was successfully utilized in a few applications. The measurement of polystyrene particle parameters gave a good agreement with analysis of the same particles with an electronic microscopy. The polystyrene particles were successfully classified into different categories from the size and refractive index map and from the SFC traces when the particles differ in size by more than 70 nm. The SFC was effectively applied to a study of dispersion radical polymerization. A study of light scattering properties of individual blood cells was carried out.

We studied the light-scattering properties of individual erythrocytes both theoretically and experimentally. Important statements relating to the optical model of a mature erythrocyte have been retrieved from theoretical simulation of light scattering of individual mature erythrocytes in different orientations with the Wentzel–Kramer–Brillouin approximation and Discrete Dipole approximation. The comparison of theoretical and experimental results has allowed us to determine the orientation effect of the hydrodynamic focussing unit of the Scanning Flow Cytometer. Moreover this study has shown a way for determination of physical characteristics of mature erythrocytes from light scattering. The possibility of measuring the light-scattering pattern of individual erythrocytes by using the SFC enables statistical analysis of the light-scattering properties of erythrocytes and could be used to study processes that result in erythrocyte lysis and erythrocyte deformability. We developed a method for simultaneous measurement of Red Blood Cell volume and hemoglobin concentration where calibration of the instrument is not necessary. The method has been tested by the measurement of sphered erythrocyte volume and hemoglobin concentration. However, the algorithm can be easily used to construct the approximating equations for other types of absorbing particles.

The performed study of light-scattering properties of individual *E. coli* cells has demonstrated the performance of SFC in measuring the light-scattering profile of cells. The measured differential cross section of *E. coli* can be the first contribution into our database of the light-scattering function of different microorganisms. With additional studies of light-scattering functions of other cells and with an enlargement of the database, microorganisms presented in the database can be identified by applying modern powerful mathematic methods and computers. Moreover, the database should consist of LSPs of cells in a logarithmic phase of growth that shows more stable LSPs of individual particles. The other important problem is determination of spheroidal modeled particle characteristics (length, diameter, and refractive index) from the LSP measured. These characteristics can be determined from the parametric solution of the inverse light-scattering problem similar to those for homogeneous spherical particles.

The Scanning Flow Cytometry allows the measurement of the distribution of cell parameters instead of measurements of mean parameters of cell populations (as is available with other light-scattering techniques). This technique seems like an ideal method for investigating individual cell-parameter changes during a cell cycle. A careful study of evolution of distributions of cell parameters, Dynamic Flow Cytometry, is a goal of our future research.

Acknowledgements

We would like to emphasize a role of Alexander Petrov, Michael Kamkha, and Erkki Soini in financial and ethical support of our young research team. Special thanks go to Professor E. Soini for initiative in the collaboration taken place in Turku University, Finland, and to Professor A. K. Petrov for support of the flow cytometric work in Novosibirsk. The authors especially acknowledge Professor S. V. Netesov and Professor V. B. Loktev (State Center for Virology and Biotechnology, Koltsovo, Novosibirsk, region) for their interest in scanning flow cytometry and the joint work. We kindly acknowledge Alfons Hoekstra from Section Computational Science, University of Amsterdam, The Netherlands, for latest fruitful collaboration and for opportunity to apply parallel computing with the University of Amsterdam super-computer.

Bibliography

Bohren, C. F. and Huffman, D. R. (1983). *Absorption and Scattering of Light by Small Particles*. Wiley, New York.

Bronk, B. V., Van De Merwe, W. P., and Stanley, M. (1992). In vivo measure of average bacterial cell size from a polrized light scattering function. *Cytometry*, 13, 155–162.

Bronk, B. V., Druger, S. D., Czege, J., and Van De Merwe, W. P. (1995). Measuring diameters of rod-shaped bacteria in vivo with polarized light scattering. *Biophysical Journal*, 69, 1170–1177.

Chernyshev, A. V., Prots, V. I., Doroshkin, A. A., and Maltsev, V. P. (1995). Measurement of scattering properties of individual particles with a scanning flow cytometer. *Applied Optics*, 34, 6301–6305.

Chernyshev, A. V., Soini, A. E., Surovtsev, I. V., Maltsev, V. P., and Soini, E. (1997). A mathematical model of dispersion radical polymerization kinetics. *Journal of Polymer Science Part B — Polymer Chemistry*, 35, 1799–1807.

Collett, E. (1993). *Polarized Light: Fundamentals and Applications*. Marcel Dekker, New York.

Cooke, D. D. and Kerker, M. (1973). Particle size distribution of colloidal suspensions by light scattering based upon single particle counts. Polystyrene latex. *J. Colloid Interface Sci.*, 42, 150–155.

Constantinides, G. N., Gintides, D., Kattis, S. E., Kiriaki, K., Paraskeva, C. A., Payatakes, A. C., Polyzos, D., Tsinopoulos, S. V., and Yannopoulos, S. N. (1998). Computation of light-scattering by axisymmetrical nonspherical particles and comparison with experimental results. *Applied Optics*, 37, 7310–7319.

Damaschke, N., Gouesbet, G., Grehan, G., and Tropea, C. (1998). Optical technique for the characterization of non-spherical and non-homogeneous particles. *Measurement Science and Technology*, 9, 137–140.

de Grooth, B. G., Terstappen, L. W. M. M., Puppels, G. J., and Greve, J. (1987). Light-scattering polarization measurements as a new parameter in flow cytometry. *Cytometry*, 8, 539–544.

Doornbos, R. M. P., Schaeffer, M., Hoekstra, A. G., Sloot, P. M. A., de Grooth, B. G., and Greve, J. (1996). Elastic light-scattering measurements of single biological cells in an optical trap. *Applied Optics*, 35, 729–734.

Draine, B. T. (1988). The discrete-dipole approximation and its application to interstellar graphite grains. *Astrophys. J.*, 333, 848–872.

Draine, B. T. and Goodman, J. (1993). Beyond Clausius–Mossotti: wave propagation on a polarizable point lattice and the discrete dipole approximation. *Astrophys. J.*, 405, 685–697.

Draine, B. T. and Flatau, P. J. (1994). Discrete-dipole approximation for scattering calculations. *J. Opt. Soc. Am. A*, 11, 1491–1499.

Ellis, R. W. and Sobanski, M. A. (2000). Diagnostic particle agglutination using ultrasound: a new technology to rejuvenate old microbiological methods. *J. Med. Microbiol.*, 49, 853–859.

Frengen, J., Lindmo, T., Paus, E., Schmid, R., and Nustad, K. (1995). Dual analyte assay based on particle types of different size measured by flow cytometry. *Journal of Immunological Methods*, 178, 141–151.

Freund, R. W. and Nachtigal, N. M. (1991). QMR: a quasi-minimal residual method for non-Hermitian linear systems. *Numer. Math.*, 60, 315–339.

Fung, Y. C., Tsang, W. C., and Patitucci, P. (1981). High-resolution data on the geometry of red blood cells. *Biorheology*, 18, 369–385.

Godefroy, C. and Adjouadi, M. (2000). Particle sizing in a flow environment using light scattering patterns. *Journal of Particles and Particle Systems Characterization*, 17, 47–55.

Goedecke, G. H. and O'Brien, S. G. (1988). Scattering by irregular inhomogeneous particles via the digitized Green's function algorithm. *Appl. Opt.*, 27, 2431–2438.

Goodman, J. J., Draine, B. T., and Flatau, P. J. (1991). Application of fast-Fourier-transform techniques to the discrete-dipole approximation. *Opt. Lett.*, 16, 1198–1200.

Hammer, M., Schweitzer, D., Michel, B., Thamm, E., and Kolb, A. (1998). Single scattering by red blood cells. *Applied Optics*, 37, 7410–7418.

Hirst, E. and Kaye, P. H. (1996). Experimental and theoretical light scattering profiles from spherical and nonspherical particles. *Journal of Geophysical Research*, 101, 19231–19235.

Hoekstra, A. G. and Sloot, P. M. A. (1993). Dipolar unit size in coupled-dipole calculations of the scattering matrix elements. *Optics Letters*, 18, 1211–1213.

Hoekstra, A. G., Grimminck, M. D., and Sloot, P. M. A. (1998). Large scale simulation of elastic light scattering by a fast discrete dipole approximation. *Int. J. Mod. Phys. C*, 9, 87–102.

Jackson, J. D. (1975). *Classical Electromagnetism*. Wiley, New York.

Jain, A. K., Duin, R. P. W., and Mao, J. (2000). Statistical pattern recognition: a review. *IEEE Transactions on Pattern Analysis and Machine Intelligence*, 22, 4–37.

Jones, M. R., Curry, B. P., Brewster, M. Q., and Leong, K. H. (1994). Inversion of light-scattering measurements for particle size and optical constants — theoretical study. *Applied Optics*, 33, 4025–4034.

Kachel, V., Fellner-Feldegg, H., and Menke, E. (1990). Hydrodynamic properties of flow cytometry instruments. In: *Flow Cytometry and Sorting*. Melamed, M. R., Lindmo T., and Medelsohn, M. L. (Eds). 2nd ed. John Wiley and Sons Inc., New York.

Kaye, P. H., Hirst, E., and Wang-Thomas, Z. (1997). Neural-network-based spatial light-scattering instrument for hazardous airborne fiber detection. *Applied Optics*, 36, 6149–6156.

Kaye, P. H. (1998). Spatial light-scattering analysis as a means of characterizing and classifying non-spherical particles. *Measurement Science and Technology*, 9, 141–149.

Kerker, M. (1969). *The Scattering of Light and other Electromagnetic Radiation.* Academic, New York.

Kim, Y. R. and Ornstein, L. (1983). Isovolumetric sphering of erythrocytes for more accurate and precise cell volume measurement by flow cytometry. *Cytometry*, 3, 419–427.

Klett, J. D. and Sutherland, R. A. (1992). Approximate methods for modeling the scattering properties of nonspherical particles: evaluation of the Wentzel-Kramers-Brillouin method. *Applied Optics*, 31, 373–386.

Krossbacher, L. M., Mima J., and Psenner R. (1998). Determination of bacterial cell dry mass by transmission electron microscopy and densitometric image analysis. *Appl. Environ Microbiol.*, 64, 688–694.

Lindmo, T., Bormer, O., Ugelstad, J., and Nustad, K. (1990). Immunometric assay by flow cytometry using mixtures of two particle types of different affinity. *Journal of Immunological Methods*, 126, 183–189.

Loken, M. R., Sweet, R. G., and Herzenberg, L. A. (1976). Cell discrimination by multiangle light scattering. *J. Histochem. Cytochem.*, 24, 284–291.

Long, M., Goldsmith, H. L., Tees, D., and Zhu, C. (1999). Probabilistic modeling of shear-induced formation and breakage of doublets cross-linked by receptors-ligand bonds. *Biophys. J.*, 76, 1112–1128.

Lopatin, V. N. and Shepilevich, N. S. (1996). Consequences of the integral wave equation in the Wentzel-Kramers-Brillouin approximation. *Optics and Spectroscopy*, 81, 103–106.

Ludlow, I. K. and Kaye, P. H. (1979). A scanning diffractometer for the rapid analysis of microparticles and biological cells. *Journal of Colloid and Interface Science*, 69, 571–589.

Ludlow, I. K. and Everitt, J. (1995). Application of Gegenbauer analysis to light scattering from spheres. *Physical Review E*, 5, 2516–2526.

Maltsev, V. P. and Chernyshev. US Patent Number 5,650,847. Date of patent: Jul. 22.

Maltsev, V. P., Chernyshev, A. V., Semyanov, K. A., and Soini, E. (1996). Absolute real-time measurement of particle size distribution with the flying light-scattering indicatrix method. *Applied Optics*, 35, 3275–3280.

Maltsev, V. P. and Lopatin, V. N. (1997). A parametric solution of the inverse light-scattering problem for individual spherical particles. *Applied Optics*, 36, 6102–6108.

Maltsev, V. P. (1994). Estimation of morphological characteristics of single particles from light scattering data in flow cytometry. *Russian Chemical Bulletin*, 43, 1115–1124.

Maltsev, V. P. (2000). Scanning flow cytometry for individual particle analysis. *Review of Scientific Instruments*, 71, 243–255.

Marx, E. and Mulholland, G. W. (1983). Size and refractive index determination of single polystyrene spheres. *J. of the National Bureau of Standards*, 88, 321–338.

Mazeron, P., Muller, S., and Azouzi, H. El. (1997). Deformation of erythrocytes under shear: a small-angle light scattering study. *Biorheology*, 34, 99–110.

Min, S. and Gomez, A. (1996). High-resolution size measurement of single spherical particles with a fast Fourier transform of the angular scattering intensity. *Applied Optics*, 35, 4919–4926.

Mishchenko, M. I., Hovenier, J. W., and Travis, L. D. (Eds) (2000). *Light Scattering by Nonspherical Particles, Theory, Measurements, and Applications*. Academic Press.

Neukammer, J., Gohlke, C., Höpe, A., Wessel, T., and Rinneberg, H. (2003). Angular distribution of light scattered by single biological cells and oriented particle agglomerates. *Applied Optics*, 42, 6388–6397.

Newman, D. J., Henneberry, H., and Price, C. P. (1992). Particle enhanced light scattering immunoassay. *Annals of Clinical Biochemistry*, 29, 22–42.

Nilsson, A. M. K., Alsholm, P., Karlsson, A., and Andersson-Engels, S. (1998). T-matrix computations of light scattering by red blood cells. *Applied Optics*, 37, 2735–2748.

Patitsas, A. J. (1973). A simple method for determining the size of a sphere from extrema of the scattering intensity I. Dielectric sphere. *J. Colloid Interface Science*, 45, 359–371.

Peterson, A. F., Ray, S. L., Chan, C. H., and Mittra, R. (1991). Numerical implementation of the conjugate gradient method and the CG-FFT for electromagnetic scattering. In: *Application of Conjugate Gradient Method to Electromagnetics and Signal Processing.* Sarkar, T. K. (Ed). Chap. 5. Elsevier, New York,

Peula-Garsia, J. M., Bolivar, J. A. M., Velasco, J., Rojas, A., and Gonzales, F. G. (2002). Interaction of bacterial endotoxine (lipopolysaccharide) with latex particles: application to latex agglutination immunoassays. *J. olloid Interface Sci.*, 245, 230–236.

Press, W. H., Flannery, B. P., Teukolsky, S. A., and Vetterling, W. T. (1990). *Numerical Recipes in Pascal.* Cambridge University Press, Cambridge.

Purcel, E. M. and Pennypacker, C. R. (1973). Scattering and absorption of light by nonspherical dielectric grains. *Astrophys. J.*, 186, 705–714.

Quist, G. M. and Wyatt, P. J. (1985). Empirical solution to the invers-scattering problem by the optical strip-map technique. *J. Opt. Soc. Am. A*, 2, 1979–1985.

Sarkar, T. K., Yang, X., and Arvas, E. (1988). A limited survey of various conjugate gradient methods for complex matrix equations arising in electromagnetic wave interactions. *Wave Motion*, 10, 527–546.

Simon, H. (1999) *Neural Networks A Comprehensive Foundation.* 2nd Edition. Prentice-Hall, Upper Saddle River, New Jersey.

Singham, S. B. and Bohren, C. F. (1987). Light scattering by an arbitrary particle: a physical reformulation of the coupled-dipoles method. *Opt. Lett.*, 12, 10–12.

Singham, S. B. and Salzman, G. C. (1986). Evaluation of the scattering matrix of an arbitrary particle using the coupled dipole approximation. *J. Chem. Phys.*, 84, 2658–2667.

Skalak, R., Tozeren, A., Zarda, R. P., and Chien, S. (1973). Strain energy function of red blood cell membranes. *Biophysical J.*, 13, 245–264.

Soini, J. T., Chernyshev, A. V., Hanninen, P. E., Soini, E., and Maltsev, V. P. (1998). A new design of the flow cuvette and optical set-up for the scanning flow cytometer. *Cytometry*, 31, 78–84.

Terstappen, L. W. M. M., de Grooth, B. G., Visscher, K., van Kouterik, F. A., and Greve, J. (1987). Four-parameter white blood cell differential counting based on light scattering measurements. *Cytometry*, 9, 39–43.

Terstappen, L. W. M. M., Johnson, D., Mickaels, R. A., Chen, J., Olds, G., Hawkins, J. T., Loken, M. R., and Levin, J. (1991). Multidimensional flow cytometric blood cell differentiation without erythrocyte lysis. *Blood Cells*, 17, 585–602.

Tsinopoulos, S. V. and Polyzos, D. (1999). Scattering of He–Ne-laser light by an average-sized red-blood-cell. *Applied Optics*, 38, 5499–5510.

Tycko, D. H., Metz, M. H., Epstein, E. A., and Grinbaum, A. (1985). Flow-cytometric light scattering measurement of red blood cell volume and hemoglobin concentration. *Applied Optics*, 24, 1355–1365.

Ulanowski, Z., Wang, Z., Kaye, P. H., and Ludlow I. K. (1998). Application of neural networks to the inverse light scattering problem for spheres. *Applied Optics*, 37, 4027–4033.

van de Hulst, H. C. (1957) *Light Scattering by Small Particles*. Wiley, New York.

van Oss, C. J. (2000). Precipitation and agglutination. *J. Immunoassay*, 21, 143–164.

Wang, Z., Ulanowski, Z., and Kaye, P. H. (1999). On solving the inverse scattering problem with RBF neural networks: Noise-free case. *Neural Comp. Appl.*, 8, 177–186.

Waterman, P. C. (1971). Symmetry, unitarity, and geometry in elecreomagnetic scattering. *Phys. Rev. D*, 3, 825–839.

Wielaard, D. J., Mishchenko, M. I., Macke, A., and Carlson, B. E. (1997). Improved T-matrix computations for large, nonabsorbing and weakly absorbing nonspherical particles and comparison with geometric optics approximation. *Applied Optics*, 36, 4305–4313.

Wyatt, P. J. (1972). Light scattering in the microbial world. *Journal Colloid and Interface Science*, 39, 479–491.

Yang, W. (2003). Improved recursive algorithm for light scattering by a multilayered sphere. *Applied Optics*, 42, 1710–1720.